Computational Structural Behavior

Cheng Wang

Computational Structural Behavior

Springer

Cheng Wang
Tongji University
Shanghai, China

ISBN 978-981-95-1012-2 ISBN 978-981-95-1013-9 (eBook)
https://doi.org/10.1007/978-981-95-1013-9

This Springer imprint is published by the registered company Springer Nature Singapore Pte Ltd.
The registered company address is: 152 Beach Road, #21-01/04 Gateway East, Singapore 189721, Singapore

Preface

Why do humans, animals, and even evolving systems behave in predictable ways? Research reveals these behaviors are not random. They follow hidden "blueprints"—rules governing timing, spatial patterns, and logic. Just as materials scientists study atoms to explain why steel is strong, we can decode behavior by analyzing its structure. This book introduces Computational Structural Behavior—a mathematical framework to model behavior. Like quantum theory revolutionized materials science, this approach may transform how we understand cognition and design AI.

By decoding these structural rules, we propose a groundbreaking framework: Computational Structural Behavior. This system aims to establish new theoretical foundations and methodological systems for behavioral modeling and prediction. Just as the quantum theory revolution propelled a paradigm shift in materials science, the establishment of behavioral structurology may catalyze a new wave of breakthroughs in cognitive science and intelligent technologies.

This book is constructed upon two core objectives: first, to provide a systematic theoretical reference for behavioral science research; second, to develop a comprehensive methodological textbook for graduate students and researchers in related fields. The content architecture follows a progressive logic from theory to application: beginning with in-depth analysis of fundamental principles and mathematical models of behavioral structure, then detailing structural computation methods based on modern machine learning algorithms such as graph computation. The novel Behavioral Molecular Graph is a method to break behaviors into interconnected units, akin to how molecules form materials. It not only captures the essential structural features of behavior but also enables efficient computational analysis through advanced techniques such as graph neural networks. It can offer a practical and scalable solution to key challenges in modern behavioral research.

Researchers across fields urgently require a unified language to model behavior—a need that drives this work. Such a language must: (1) accurately describe behavioral structural characteristics across different scales and domains; (2) support complete translation from theoretical models to engineering implementations; (3) adapt to diverse research requirements and application scenarios. We anticipate that this research will not only deepen theoretical understanding of computational behavioral structures but also offer tools to solve real-world problems—for example, predicting crowd movements during disasters or optimizing team decision-making in corporations.

Shanghai, China Cheng Wang

Declarations

Competing Interests The author has no competing interests to declare that are relevant to the content of this manuscript.

Contents

Chapter 1
Overview of Computational Structural Behavior

Abstract Efforts to comprehensively describe behavior have advanced representation-based methods, particularly those using machine intelligence models like deep learning models, to capture behavioral information. As behavior complexity increases, the expressive power of these intelligence models reaches a bottleneck. Behavior possesses inherent structural information, which allows for efficient description of increasingly complex behaviors. To concretize behavior structure into a format suitable for processing, this book adopts graphs to reflect the relationships between behavioral attributes at a fine-grained level and designs the behavioral molecular structure model. This book designs the different representative downstream tasks to test the performance of the model on renowned public datasets. The model has high performance in detecting behavior, predicting behavior, and generating behavior. Overall, computational structural behavior is a framework that deciphers behavioral patterns with precision akin to engineering analysis at the atomic level and offers a high-performance and expressive approach for behavioral computing research.

1.1 Introduction to Computational Structural Behavior

Behavior refers to any observable and measurable action, response, or activity [1]. When we observe and measure behavior, there is often a significant gap between our expectations of its expressive power and the current behavior modeling paradigm's ability to capture behavioral expression [2]. A complete and comprehensive description of behavior will require large amounts of data, such as high-resolution video clips, to capture the full range of actions, responses, and activities involved [3]. Processing and analyzing such behavioral data are undoubtedly computationally intensive [4]. Reducing such behavioral descriptions into more expressive observational features has significant practical importance [5]. Researchers have opened up a new pathway for behavioral expression by utilizing machine intelligence models to gain a deep understanding of the movement of the objective world. Going even further, it is a triumph that simplifies behavioral descriptions into low-dimensional

C. Wang, *Computational Structural Behavior*,
https://doi.org/10.1007/978-981-95-1013-9_1

representations for just limited degrees of freedom using representation-based methods [6–8].

From human social activities to animal movements in the biological world to airflow in the natural world, all descriptions of behavior are subject to limitations of predictability and diversity [9–13]. When we talk about behavior, we usually assume that individual behavior is not entirely random and is predictable to some extent [14–17]. However, from the perspective of an external observer, the behavior seems to be relatively random (i.e., there are differences between behaviors) compared to the behavior of others [14, 18–22]. So, it is possible to develop high-expressive models of behavior under certain conditions, such as detecting malicious behavior [23–25], predicting behavior trajectories [26–28], generating simulated behavior [29–31], etc.

Various methods can be employed to provide high-precision observational data for describing behavior. After decades of research, representation-based methods, utilizing the nonlinear transformation ability of deep learning, have pushed the understanding of behavior to a higher level [32–34]. Behavioral representations have greater expressive power than direct observation of behavior. From the beginning of association within behavior sequences to association in behavior communities, increasingly complex behavior models have been developed. They quantify the regularity of behavior using representations and cluster or separate them in low-dimensional space. Unfortunately, as the complexity of behavior increases, the expressive power of these models reaches a bottleneck. Further improvement in behavioral expression is undoubtedly a persistent challenge in the scientific community.

Understanding behavior, especially studying the correlation between behaviors, is a sophisticated issue. Facing two behaviors, one qualitative answer is to compare and discover their differences and predictability by their consistency/inconsistency. Just as in genetics, we analyze the DNA structure to achieve a refined quantitative analysis (which can be seen as a phenomenological description of genetic patterns) to explain why we resemble our parents or why we differ in appearance from others. Behavior can have a multi-level descriptive structure that is able to encode information. In broad contexts, the structure of behavior is a candidate for a unified behavioral description. How can we construct a complex structure of behavior to further enhance our ability to express behavior? All of these puzzles point to the problem of defining the behavioral structure. In mathematics, the structure of a mathematical object refers to a non-empty basic set A (the research object) and a set of relationships R that satisfy certain properties defined on the basic set A [35–38].

Taking inspiration from the interactions between atoms in molecular structures [39], we simulate internal correlations in behavior. Because this phenomenon is not widely known, we coin the term *behavioral molecular structure* to facilitate research in this field. Our research is driven by the observation that, in the material world, molecules have structures that can be described by chemical bonds representing strong interactions between neighboring atoms (or ions) and fine-grained atoms [40–42]. Similarly, we have reason to believe that behavior, as an observable objective

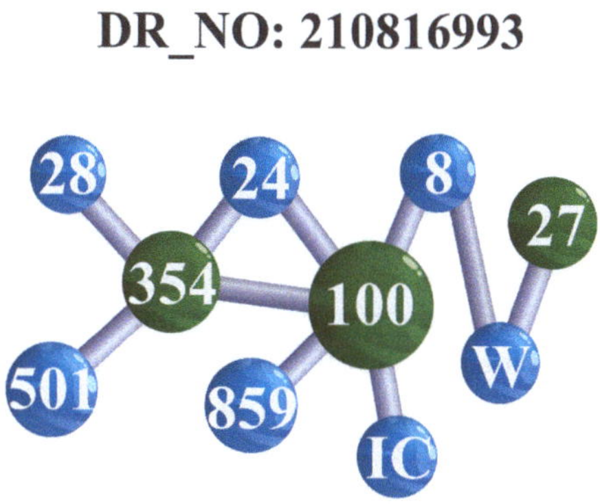

Fig. 1.1 Illustration of behavioral molecular structure

reality, has a measurable structure that can be explored and analyzed using a variety of tools and techniques. Basic scientific research (e.g., mathematical hypotheses used in physics) on the material world can provide insights into the behavior under complex conditions. By uncovering this underlying mechanism (like structure), researchers can gain new insights into the mechanisms that govern behavior, leading to more accurate and sophisticated models of behavior. This, in turn, can help to inform the development of more effective behavior models for a wide range of subtasks.

To work toward a sharper definition, consider the case in which the behavior is limited behavioral attributes, all the behavioral attributes consist of the basic set A. Furthermore, we obtain the set of relationships R after determining the relationships between behavioral attributes. Thus, we can concretize observable behavior as a structure defined on sets A and R (as shown in Fig. 1.1). Here, the behavior is visualized as a structure composed of nodes and edges, where we select ten behavioral attributes for the criminal behavior (numbered 210816993).

Identifying key structural features is crucial for revealing molecular activities and properties, and the same applies to behavior. The behavioral molecular structure should capture the characteristics of various behaviors. An excellent behavioral molecular structure should meet the following conditions:

- Uniqueness. Each behavior has its unique structure and properties. Even very similar behaviors may have different structures and properties due to small differences.
- Stability. Behaviors usually maintain their key structure and properties under certain conditions (such as time and place). Despite the apparent randomness of individual behavior, the daily behavior patterns hide unexpectedly high potential predictability.

1.2 From Behaviorism to Behavioral Computation

1.2.1 Behavioral Research: Paradigm Revolution

Behavior research has been a long-standing focal point in academia [5, 9, 10, 13]. Observation of behavior provides a window for understanding it [6, 17].

Behaviorist psychology in the early twentieth century attempted to reduce human behavior to controllable stimulus-response (S-R) chains in laboratory settings [43]. Watson's "Little Albert experiment" and Skinner's operant conditioning chamber established the scientific foundation for observable and quantifiable behavior [44]. However, the mechanistic theory, which views natural and social processes as interconnected systems akin to a machine, fails to account for behavioral anomalies in complex social contexts that clearly go beyond simple reward-punishment mechanisms. This contradiction catalyzed the emergence of cognitive science: Newell and Simon proposed the "Physical Symbol System Hypothesis," positing that behavior is the external projection of internal cognitive states [45].

The fundamental shift in human understanding of behavior reflects a paradigm revolution in behavioral science. Traditional economics is built on the "rational man" hypothesis, assuming individuals make optimal decisions based on complete rationality [46]. This hypothesis faced increasing challenges in the twentieth century, as interdisciplinary research in behavioral economics and psychology revealed systematic biases and irrational features in human decision-making. Kahneman and Tversky's Prospect Theory demonstrated experiments that people exhibit asymmetric risk preferences when facing gains and losses, which contradicts traditional expected utility theory [47]. Simon's theory of bounded rationality further posits that human cognitive abilities are limited and decisions are often based on the principle of satisficing rather than optimization [48]. Thaler's mental accounting theory reveals the irrational ways people categorize money [49].

The paradigm revolution in behavioral science has profoundly influenced disciplines such as economics, finance, and management. In finance, behavioral finance has provided explanations for market anomalies [50]. In management studies, behavioral decision theory has reshaped research on organizational behavior [51]. In public policy, Nudge Theory has provided new approaches to policy design [52]. This revolution has not only transformed academic research paradigms but also significantly influenced business practices and public policy formulation.

Currently, behavioral science is advancing to deep levels. The emergence of new interdisciplinary fields such as neuroeconomics and genetic economics provides novel perspectives for understanding human behavior [53, 54]. The application of big data and artificial intelligence technologies has shifted behavioral research from laboratory settings to real-world scenarios [55]. These developments bring new opportunities and challenges to behavioral science, indicating that the paradigm shift in this field will continue to deepen.

The advent of the big data era in the twenty-first century has fundamentally transformed the paradigm of behavioral research. Behavioral computing has thus entered the stage of data-intensive science. For instance, machine learning-based methods have made it possible to automatically simplify large amounts of video data into time series, using only a few behavior features with limited degrees of freedom. This approach has proven to be effective in downstream behavior recognition tasks [26, 33]. Song et al. [56] explored the predictability of human dynamics using trajectory records from millions of mobile phone users. They highlighted that with the aid of appropriate data mining algorithms, the predictability of behaviors can be transformed into practical mobility predictions. Stachl et al. [57] utilized machine learning to predict personality traits across a wide range of domains using six different types of behavioral information collected from smartphone sensors and log data.

1.2.2 Representation on Behavior: From Symbolic Logic to Neural Embedding

There has been explosive growth in behavioral data, leading to increased attention to behavioral representations [23, 25, 28]. The research on behavioral representation has evolved from symbolic logic to neural embeddings, reflecting the continuous deepening of human understanding of the nature of behavior [58].

Early behavioral representations were constrained by the framework of symbolic artificial intelligence, reducing human behavior to computable symbolic operations [59]. In the symbolic logic phase, behavior was abstracted into discrete symbols and rules [60]. Newell and Simon's Physical Symbol System Hypothesis is a representative theory of this stage, positing that intelligent behavior can be achieved through symbolic operations [45]. This approach achieved some success in fields such as expert systems but struggled to handle complex and ambiguous behavioral patterns. For instance, Terry Winograd developed the SHRDLU system, which used predicate logic to describe user operations in a virtual block world (e.g., "pick up the red block"). However, its 200 hand-coded rules were inadequate for handling the ambiguity of real-world scenarios [61, 62]. This dilemma was resolved in the twenty-first century by deep learning.

The emergence of neural embeddings marks a significant turning point in behavioral representation research. This transformation is centered on converting discrete symbolic representations into continuous vector spaces, enabling distributed representations of behavioral features [63]. At the technical implementation level, neural embeddings map behavioral features into high-dimensional vector spaces through neural networks. Taking Word2Vec as an example, its core idea is to learn word vector representations by predicting context. The Skip-gram model predicts context words from the current word, while the CBOW model predicts the current word from context words. This self-supervised learning approach allows models to automatically learn meaningful representations from large-scale behavioral data. For instance, in user behavior analysis, click sequences can be encoded as vectors to capture user preferences and behavioral patterns [64]. Behavioral features do not exist in isolation but are defined through their relationships with other features. In high-dimensional vector spaces, similar behaviors cluster together to form semantic clusters, where semantic clustering groups data points, particularly texts or documents, based on their content similarity or semantic meaning, ensuring that items within a cluster are semantically related. This representation method offers several advantages: first, it can capture subtle differences in behaviors; second, it supports vector operations, making behavioral reasoning possible; and finally, it can handle ambiguous and uncertain behavioral features.

The application of neural embeddings in behavior understanding is mainly reflected in three aspects: First, it can discover potential associations between behaviors, such as measuring behavioral similarity through vector distances. Second, it supports behavior prediction, such as predicting future behaviors based on historical behavior sequences. Third, it enables behavior clustering and classification, such as identifying abnormal behavior patterns. These applications have achieved significant success in recommendation systems, user profiling, and risk control. This phase of research has overcome the limitations of the discreteness of symbolic logic, providing richer semantic information for behavior understanding. The breakthrough of neural embedding technology lies in its resolution of the inherent problems of symbolic logic representation: the curse of dimensionality, the semantic gap, and scalability limitations. By representing behaviors as continuous vectors, it not only preserves the semantic information of behaviors but also automatically learns complex relationships between behavioral features. This technological breakthrough has laid the foundation for the subsequent application of deep learning in behavior analysis and has driven a paradigm shift in behavioral science research.

The rise of deep learning has brought revolutionary progress to the study of behavioral representation. Deep learning-based approaches have shown promise in transforming high-dimensional behavioral data into meaningful low-dimensional representations that are more compact and information-rich [1, 7]. Its core lies in automatically extracting meaningful feature representations from raw behavioral data through multi-level nonlinear transformations. It not only improves the efficiency of behavioral analysis but, more importantly, achieves a deeper understanding of complex behavioral patterns [63]. Furthermore, by learning behavior representations tailored to specific tasks, it becomes possible to better capture key behavioral patterns within the data. This, in turn, contributes to enhancing the understanding and predictive capabilities of behavioral data.

At the technical implementation level, deep learning unifies feature extraction and pattern recognition within a single framework through end-to-end learning. Convolutional neural networks (CNNs) excel in spatial feature extraction [65], while Recurrent Neural Networks (RNNs) and Long Short-Term Memory (LSTM) networks effectively handle behavioral sequence data, capturing temporal dependencies [66, 67]. The introduction of the attention mechanism allows models to focus on key behavioral features. The development of graph neural networks (GNNs) provides new tools for modeling complex behavioral relationships. These technological advancements have significantly improved the accuracy and interpretability of behavioral representations. For example, Pereira et al. [4] proposed a deep learning-based method for predicting the positions of animal body. It utilized a convolutional neural network architecture to efficiently train and evaluate images, providing probability estimates for the location of each tracked body. RNNs and their LSTM variants have advantages in time series modeling, effectively handling temporal dependencies in behavioral data. The introduction of attention mechanisms further enhances model performance, enabling dynamic focus on key behavioral features. For instance, Clipman et al. [68] used network analysis and graph neural networks, identifying key transmission nodes through attention mechanisms, providing a basis for intervention strategy formulation.

The application of deep learning methods in behavioral representation is mainly reflected in three dimensions: First, it can automatically extract multi-level behavioral features, from low-level motion features to high-level behavioral patterns. Second, it supports cross-modal behavioral analysis, such as integrating visual, auditory, and other multisource data. Third, it can model complex behavioral relationship networks, such as social interaction analysis. These applications have had a profound impact in fields such as neuroscience, epidemiology, and sociology.

The technological breakthroughs brought by deep learning are mainly reflected in four aspects: it overcomes the limitations of traditional methods that rely on manually designed features; it can handle high-dimensional, unstructured behavioral data; it supports end-to-end learning and inference; and it can model complex spatiotemporal dependencies. These breakthroughs not only improve the accuracy of behavioral analysis but, more importantly, open up new paradigms for behavioral understanding.

Currently, behavioral representation research is moving toward multimodality and explainability. Multimodal learning techniques can integrate various behavioral data such as text, images, and voice to build more comprehensive behavioral representations [69–72]. Research in Explainable Artificial Intelligence (XAI) aims to reveal the decision-making processes of deep learning models, improving model transparency [73, 74]. These advancements provide new possibilities for theoretical innovation and practical applications in behavioral science.

The evolution of behavioral representation research has not only advanced the development of artificial intelligence technologies but also deepened human understanding of the nature of behavior. From symbols to neurons, from discrete to continuous, from single to multimodal, this journey reflects a fundamental shift in scientific research paradigms and points the way for the future development of behavioral science.

1.2.3 Relation on Behavior: From Observation to Deep Mining

The associations between behaviors are complex and multidimensional, reflecting not only in the sequential patterns of individual behaviors but also in the interaction networks of group behaviors [75, 76]. Understanding these associations is crucial for predicting behaviors, intervening in behaviors, and designing behavior-oriented systems. Behavioral association research has formed a multidimensional, multi-level research system. On one hand, it covers various levels, roughly divided into individual behavior associations and group interaction associations; on the other hand, it can be divided into temporal and spatial dimensions in terms of association dimensions. Significant progress has been made in both theoretical methods and practical applications, but many challenges remain.

In individual behavior research, fine-grained behavior analysis based on deep learning has become mainstream. In the temporal dimension, deep learning models such as LSTM and transformer have significantly enhanced the ability to analyze behavioral sequences [27, 30, 66, 75]. LSTM captures long-term dependencies in behavioral sequences through its gating mechanism, excelling in processing continuous behavioral data. The transformer model, utilizing self-attention mechanisms, can process sequence data in parallel and has advantages in capturing complex temporal patterns. However, these models still face limitations when dealing with sparse behaviors and rare events. For example, in healthcare, analyzing patients' irregular medication behaviors, or in finance, detecting anomalous transaction behaviors, existing models often struggle to accurately identify and predict these behaviors. In the spatial dimension, advancements in computer vision have provided powerful tools for analyzing spatial features of behaviors [4, 7]. Deep learning-based behavior recognition algorithms can precisely quantify spatial features of behavior patterns, such as posture and motion trajectories. The development of 3D convolutional neural networks and pose estimation algorithms has made the analysis of complex spatial behaviors possible.

Group behavior research focuses on the modeling and analysis of social networks. The development of GNNs has provided new tools for modeling complex social relationship networks, yet the emergent features—new properties or behaviors arising from interactions among basic components within a system—remain challenging. These emergent features, which cannot be directly inferred by analyzing individual components in isolation, often manifest as enhanced system functions, self-organizing structures, or intelligent behaviors, particularly in complex systems like neural networks, social networks, or multi-agent systems. In the application of GNNs, researchers have developed various innovative methods to address the challenges of group behavior modeling [77–79]. Graph Attention Networks introduce attention mechanisms to dynamically capture the intensity of interactions among group members, excelling in identifying key nodes and predicting information propagation paths [80]. Spatio-Temporal Graph Convolutional Networks integrate spatiotemporal information, enabling the modeling of dynamic evolution processes in group behaviors [81]. Breakthroughs in group behavior research will bring revolutionary impacts to multiple fields. In urban planning, accurate group behavior prediction will optimize urban spatial layouts; in public health, improved disease transmission models will enhance epidemic control efficiency; in financial regulation, group behavior analysis will help prevent systemic risks. These application prospects highlight the significant value of group behavior research.

Despite significant progress in theory and methodology, behavioral association research still faces substantial limitations that constrain the depth of research and the full realization of its application value.

At the data level, behavioral association research faces triple challenges of data quality, diversity, and availability. The collection of behavioral data often involves noise and bias, such as sensor errors and sampling bias, affecting data reliability. A single data source is insufficient to comprehensively reflect behavioral features, while integrating multisource data faces issues like inconsistent formats and temporal asynchrony. Meanwhile, privacy protection and data security requirements limit data sharing and usage, preventing many valuable behavioral datasets from being fully utilized.

At the information level, behavioral association research faces challenges in information extraction and representation. The information embedded in behavioral data is often implicit, high-dimensional, and unstructured, making it challenging to extract meaningful behavioral features. Existing feature engineering methods rely on domain knowledge and struggle to adapt to complex behavioral patterns. Additionally, the associations between different behavioral features may be nonlinear and dynamically changing, making traditional linear analysis methods inadequate for capturing these complex relationships.

At the model level, behavioral association research faces limitations in model expressiveness, interpretability, and generalization capabilities. While deep learning models can capture complex behavioral patterns, they often lack interpretability, making it difficult to reveal the intrinsic mechanisms of behavioral associations. Existing causal inference methods have limited effectiveness when dealing with high-dimensional behavioral data, struggling to distinguish between correlations and causal relationships. Moreover, models that perform well on specific datasets often experience performance degradation when transferred to new scenarios, indicating insufficient generalization capabilities.

These limitations interact in complex ways. Data quality issues affect the effectiveness of information extraction, inadequate information representation restricts model expressiveness, and model limitations, in turn, influence data utilization methods. For example, privacy protection requirements limit data sharing, leading to insufficient training data for models and thereby affecting their generalization capabilities.

1.2.4 Structure in Behavior: Microscopic Analysis of Complex Behavior

Traditionally, existing works rely on macrolevel perspectives to conceptualize potential behavioral mechanisms. Typically, complex behaviors are transformed into quantifiable features or representations to reveal interrelations between behaviors. In contrast, this book focuses on studying the micro-mechanisms of behavior, proposing a way that emphasizes the interactions between internal attributes of behavior to explain behavioral mechanisms at the microlevel [77, 82, 83]. Delving into the internal structure of behavior holds significant theoretical and practical importance. Theoretically, it helps reveal the mechanisms and laws governing behavior generation, providing a new research paradigm for behavioral science. Practically, understanding behavioral structures can offer scientific foundations for behavior prediction, intervention, and design, such as in personalized education and precision medicine.

Structure refers to the organization and interrelationships between the components of a system. For behavior, structure implies the constituent units of behavior and their interaction patterns. This structure is not visually apparent but requires microlevel analysis to uncover inherent laws. As an objective existence, the structural characteristics of behavior are necessarily reflected in the correlations between its internal attributes. These correlations are not random but follow specific organizational principles, reflecting the functional requirements and generative mechanisms of behavior. The correlations between behavioral attributes can be understood across multiple dimensions. Temporally, behavior manifests as ordered action sequences with causal relationships and temporal dependencies between consecutive actions. Spatially, behavior involves coordinated movements of body parts, with spatial correlations between different body parts. Cognitively, perception, decision-making, and execution processes influence each other, forming complex cognitive correlation networks.

The main challenges in behavioral structure research include defining the basic units of behavior, quantifying the relationships between behavioral units, and establishing links between behavioral structure and function. Molecular science provides important insights for behavioral structure research [39, 84]. Just as molecules are composed of atoms connected by chemical bonds, behavior can be viewed as composed of basic units connected in some manner. This analogy inspires the theoretical framework of behavioral molecular structure for parsing and modeling complex behaviors. The core idea is to treat complex behaviors as molecules and decompose them into minimal constituent units behavioral atoms and by studying the interactions between these atoms build a systematic understanding of behavior.

In biology, molecules can be neurotransmitters (chemical messengers that transmit signals across synapses in the nervous system), hormones (signaling molecules produced by glands that regulate various physiological processes), receptors (protein molecules on cell surfaces or inside cells that bind to specific molecules like neurotransmitters or hormones to initiate cellular responses), etc. For example, dopamine is a neurotransmitter related to reward and motivation, and its release levels and receptor-binding capacity directly influence individual behavior. Similarly, behavioral atoms are the minimal units or basic components constituting behavioral molecules. These behavioral atoms can be concrete physical entities (e.g., neurotransmitters in organisms or data packets in computer networks) or abstract concepts (e.g., user behavior patterns or rules in social interactions). The characteristics of behavioral atoms and their relationships with other behavioral atoms determine their role and significance in behavioral molecules. For example, in computer networks, behavioral atoms can be data packets, protocols, sessions, etc. Here, data packets are the minimal units of network communication, and their header and payload information determine the specific manifestations of network behavior.

Metadata and contextual information provide critical supplementary dimensions for behavioral analysis, transforming the understanding of behavioral atoms from isolated events to systematic processes. This shift not only enriches behavioral analysis but, more importantly, enhances its accuracy and explanatory power. In biology, the study of neurotransmitters fully demonstrates the importance of metadata. The release of neurotransmitters depends not only on their type and quantity but also critically on the temporal and spatial patterns of release. For example, the function of dopamine in the reward system is closely related to its spatiotemporal release characteristics. Accurately recording this metadata, combined with contextual information about neural activity, enables researchers to reconstruct functional maps of neural circuits and reveal the neural mechanisms underlying behavior.

In network traffic behavior, metadata of data packets provides the foundation for network behavior analysis. Timestamp information can reveal temporal patterns of network traffic for detecting anomalies; source and destination address information can reconstruct network topology and analyze information propagation paths. Combined with contextual information about network states, such as routing tables and bandwidth utilization, it enables more accurate diagnosis of network issues and optimization of network performance.

Achieving the transition from behavioral representation to structural computation requires addressing several technical challenges: first, defining appropriate behavioral atoms and interaction relationships; second, developing machine learning algorithms capable of processing behavioral molecular structure data; and third, establishing application models linking behavioral structure and function. Solving these issues will propel behavioral science research into a new developmental stage.

Complex behaviors are not merely simple collections of behavioral atoms but result from intricate interactions among these atoms. The interactions between behavioral atoms can be direct (e.g., neurotransmitter-receptor binding) or indirect (e.g., associations between user behavior patterns). By studying these interactions, macroscopic models of behavior can be constructed. Formalizing behavioral molecular structures as graph structures provides powerful mathematical tools and computational frameworks for behavioral analysis. This formalization not only enables the visualization of behavioral structures but, more importantly, supports the computation and analysis of complex behavioral patterns.

In graph structure representations, behavioral atoms are nodes, and the interaction relationships between behavioral atoms are edges. This representation naturally captures the compositional relationships and dynamic features of behavior. Direct interactions correspond to strongly connected edges, reflecting immediate influences between behavioral atoms; indirect interactions are represented through path connections, revealing cascading effects and long-range associations in behavior.

The rise of GNNs has provided new technical pathways for processing behavioral molecular structure data. GNNs can directly handle graph-structured data, learning feature representations of nodes and edges to capture the topological information and dynamic characteristics of behavioral molecular structures, bringing revolutionary methodological innovations to behavioral analysis.

GNNs exhibit unique advantages when processing behavioral molecular structure data: First, GNNs aggregate neighbor information through message-passing mechanisms, naturally capturing local interactions between behavioral atoms. Second, GNNs support multi-level feature learning, enabling the modeling of both local structures and global patterns of behavioral molecules. Third, GNNs can handle heterogeneous graph data, accommodating different types of behavioral atoms and interaction relationships. For example, in complex skill analysis, GNNs can simultaneously model spatiotemporal relationships between action units and functional connections between cognitive processes.

Establishing application models linking behavioral structure and function requires deep integration of behavioral analysis theory with specific application scenarios. For instance, in cybersecurity and network services, such modeling can significantly enhance system performance and security.

Building behavior structure-function models involves addressing several technical challenges: designing appropriate graph structures to accurately capture key behavioral features in application scenarios, developing efficient graph learning algorithms capable of processing large-scale dynamic graph data, establishing reliable evaluation systems to validate model performance in real-world applications, and considering the real-time and scalability requirements of models to meet deployment needs.

In summary, behavioral molecular structure is a theoretical framework that decomposes complex behaviors into minimal units and studies their interactions. Through this framework, researchers can gain deeper insights into the nature of behavior and provide scientific foundations for its analysis and prediction. Whether in biology, computer networks, social sciences, or artificial intelligence, research on behavioral molecular structures holds a significant theoretical and practical value. As technology advances and interdisciplinary research deepens, the concept of behavioral molecular structure will continue to offer new perspectives and methods for parsing and modeling complex behaviors.

1.3 Outline of This Book

This book introduces a novel perspective for elucidating behavioral structure by capturing interactions between behavioral attributes. Specifically, a tangible graph structure is proposed for behavior description, reflecting the uniqueness and stability of behaviors. To achieve this, the behavioral molecular structure model is designed to generate attribute representations rich in behavioral information. In this model, attribute interactions are analogized to atomic bonds, forming a comprehensive set of relationships to describe behavior structure. Graph representation learning algorithms are then employed to extract rich behavioral information, serving as the foundation for constructing highly expressive behavior models. The effectiveness of the proposed model is validated across different representative downstream tasks: behavior detection, behavior prediction, and behavior generation. Experimental results demonstrate that utilizing atomic-level attribute interactions in behavior significantly improves performance on these tasks. Furthermore, an initial explanation for the observed performance gains is provided from an expressive power perspective. Overall, behavior molecular structure offers a framework for expressing behavioral structure, aiming to propose promising signals for the deep utilization of behavioral information. It encourages increased awareness within the research community and lays the groundwork for future studies on behavioral structure.

Next, this book focuses on introducing some representative technologies related to behavior structure and explores the relevant key applications in behavior computing. The content structure of this book is shown in Fig. 1.2.

This book begins with an introduction to behavioral molecular structure as a framework for enhancing behavior expression.

- Enhancing Behavior Expression Through Behavioral Structure (Chap. 2).
  Next, customized structure-based models are designed to generate and represent behavioral structure, further improving the performance of behavioral models.
- A novel motif-based behavior reconstruction model, MORE, reconstructs behavior as frequent connectivity patterns at the attribute level rather than representing behaviors directly (Chap. 3).

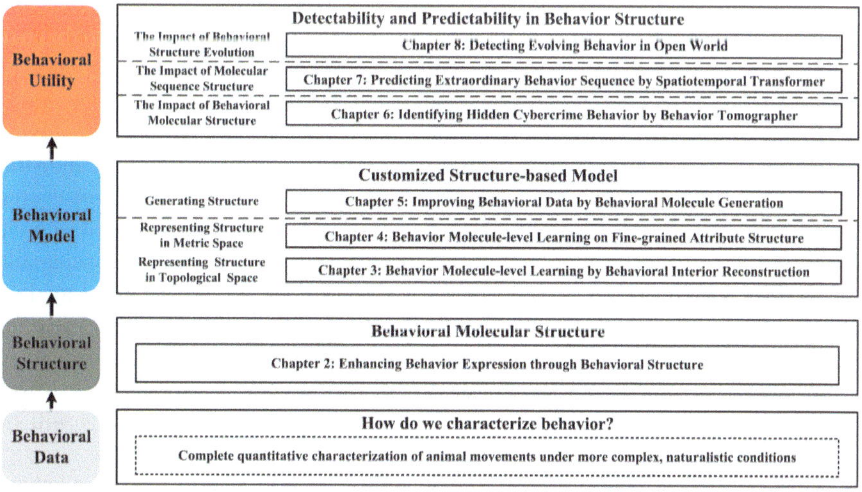

Fig. 1.2 Architecture of this book

- A tabular data representation model, the behavior molecule model, maximizes structural information gain through graph-level molecular learning (Chap. 4).
- A generative model for tabular data, TabBMG, integrates inter-attribute relationships into the tabular data generation process (Chap. 5).

  Finally, tailored strategies address typical challenges in behavior detection and prediction, demonstrating how behavioral structure enhances model effectiveness.
- Behavior tomographer (BT) adopts fine-grained detectors to reflect interior associations within behaviors. Its core idea is to reconstruct interior information about behavior structure from multiple slices, much like computed tomography in modern medicine, which enables the reconstruction of the internal body (Chap. 6).
- X-Trafformer, a graph spatiotemporal transformer model, transforms traffic sequences into generalized traffic behaviors, enabling multivariable prediction across networks by leveraging attribute matching (Chap. 7).
- EvoFD, an online evolving fraud detection framework, employs continual learning to weaken the bias caused by the anchoring effect on the learned information, adapting to dynamic fraud patterns (Chap. 8).

  The key technologies and methods of computational structural behavior include the following aspects:
- Behavioral molecular structure. The behavioral space is close to infinity, but it can be limited by decomposing behavior into a combination of finite behavioral attributes (just like atoms in molecular structure). In this way, the behavioral expression is transformed into different combinations of behavioral attributes. A behavioral attribute (like a single atom) is influenced by other behaviors, and perturbations to the behavioral molecular structure can lead to huge differences between behaviors. This book introduces a general solution, the behavioral molecular structure (BMS) model. BMS characterizes behaviors at the atomic level, with the behavioral attributes that make up the behavior being likened to atoms, where the interrelationships between attributes are studied and concretized using graphs. More details will be provided in Chap. 2.
- Behavioral interior reconstruction. Malicious behaviors have grown more sophisticated, which has rendered traditional detection methods increasingly indistinguishable. By adopting a mindset that maps behavior onto a high-dimensional attribute space, valuable clues can be derived to aid in detecting deceptive behavior by transforming nearly inseparable behavioral data into separable patterns within a behavioral attribute space. This book proposes a novel method called *mo*tif-based behavior *re*construction (*MORE*). It reconstructs behavior as the frequent connectivity pattern on the granularity of attributes instead of representing behaviors, where a transformation network is designed for learning a vector field to get expressive attribute representations. More details will be provided in Chap. 3.
- Fine-grained attribute structure. Expanding the underlying information in behavioral data presents a nontrivial task, as it frequently introduces more noise than useful information. Existing techniques primarily concentrate on treating data

as feature vectors, often disregarding the usage of fine-grained global relational information within the data. This book introduces a novel method for modeling behavior that incorporates low-noise, fine-grained, structural information. Motivated by the interatomic forces forming chemical molecules, this book captures the global relationships between attributes and models the interactions among internal attributes of behavior to construct a behavior molecule graph for each data point. This approach not only enhances the representational power of the data through structural information but also ensures the mutual isolation of behavior molecule graphs, effectively preventing the propagation of noise. More details will be provided in Chap. 4.

- Behavioral molecule generation. Challenges such as privacy concerns and information constraints result in incomplete or insufficient behavioral data, limiting the ability of models to fully capture underlying behaviors. Due to the complexity of subtle relationships between attributes in behavioral data, generating high-quality synthetic tabular data remains a significant challenge. Inspired by computer-aided drug design, this book proposes a novel generative model, TabBMG, which integrates inter-attribute relationships within tabular data. More details will be provided in Chap. 5.
- Identifying behavior by behavior tomographer. Malicious behaviors are concealed beneath benign behaviors, and the boundaries between malicious and benign behaviors in the representation space are blurred to evade mainstream deep learning-based security authentication models. This book introduces a *b*ehavior *t*omographer (BT) to reconstruct the behavior structure from three slices, agent, event, and attribute slices, enabling more granular detection of hidden cybercrimes. The core idea of BT is to reconstruct interior information about behavior structure from multiple slices, much like computed tomography in modern medicine that enables the reconstruction of the internal body. It enables the extraction of discriminative information from intricate interior associations between behavioral attributes rather than surface information meticulously crafted by criminals. Our experiments are conducted on two representative cybercrime datasets. More details will be provided in Chap. 6.
- Predicting behavior by spatiotemporal transformer. Compared to ordinary traffic prediction, these demands call for solutions to mine rich information and predict business load under broader conditions. This book proposes X-Trafformer, a graph spatiotemporal transformer model, for extraordinary traffic prediction. Unlike conventional techniques that model traffic sequences, we transform traffic sequences into traffic behaviors under generalized objects, where behaviors are initiated by generalized objects and possess specific behavioral attributes. It allows for the prediction of multiple variables in traffic data across different networks and services, leveraging the matching of behavioral attributes among behavior objects and events. X-Trafformer incorporates multiple interrelated graph structures to capture fine-grained attribute spatiotemporal associations and coarse-grained object spatiotemporal distributions, forming the foundation for accurate prediction. More details will be provided in Chap. 7.

- Detecting evolving behavior in open world. Fraudsters can change behavior patterns frequently and at a low cost in the online space, allowing them to evade regulatory oversight. The evolving fraud is not only reflected in the drift of their superimposed risk features but also in the openness of their category. The interweaving of open category and concept drift accelerates the process of existing security methods becoming powerless. We propose an online evolving fraud detection framework, called EvoFD, to enable continual learning to cope with undercurrent surges of evolving fraud. The core idea of EvoFD is to weaken the bias caused by the *anchoring effect* on the learned information. It learns in an online streaming fashion by using instructive representations as anchors. More details will be provided in Chap. 8.

References

1. M.W. Mathis, A. Mathis, Curr. Opin. Neurobiol. **60**, 1 (2020)
2. J.A. Gallego, M.G. Perich, L.E. Miller, S.A. Solla, Neuron **94**(5), 978 (2017)
3. K. Branson, A.A. Robie, J. Bender, P. Perona, M.H. Dickinson, Nat. Methods **6**(6), 451 (2009)
4. T.D. Pereira, D.E. Aldarondo, L. Willmore, M. Kislin, S.S.H. Wang, M. Murthy, J.W. Shaevitz, Nat. Methods **16**(1), 117 (2019)
5. T.D. Sanger, J. Neurosci. **20**(3), 1066 (2000)
6. W. Bialek, Proc. Natl. Acad. Sci. **119**(18), e2021860119 (2022)
7. A. Mathis, P. Mamidanna, K.M. Cury, T. Abe, V.N. Murthy, M.W. Mathis, M. Bethge, Nat. Neurosci. **21**(9), 1281 (2018)
8. B.M. Yu, J.P. Cunningham, G. Santhanam, S. Ryu, K.V. Shenoy, M. Sahani, in *Advances in Neural Information Processing Systems*, vol. 21, pp. 1881–1888 (2008)
9. D.D.C. Teixeira, A.C. Viana, M.S. Alvim, J.M. Almeida, in *Proceedings of the 27th ACM SIGSPATIAL International Conference on Advances in Geographic Information Systems* (2019), pp. 52–61
10. J. Kulshrestha, M. Oliveira, O. Karaçalık, D. Bonnay, C. Wagner, in *Proceedings of the International AAAI Conference on Web and Social Media*, vol. 15 (2021), pp. 327–338
11. K.A. Lipinski, L.J. Barber, M.N. Davies, M. Ashenden, A. Sottoriva, M. Gerlinger, Trends in Cancer **2**(1), 49 (2016)
12. E. Guilyardi, W. Cai, M. Collins, A. Fedorov, F.F. Jin, A. Kumar, D.Z. Sun, A. Wittenberg, Bull. Am. Meteorol. Soc. **93**(2), 235 (2012)
13. V. Colizza, A. Barrat, M. Barthélemy, A. Vespignani, Proc. Natl. Acad. Sci. **103**(7), 2015 (2006)
14. C. Song, Z. Qu, N. Blumm, A.L. Barabási, Science **327**(5968), 1018 (2010). https://doi.org/10.1126/science.1177170. https://www.science.org/doi/abs/10.1126/science.1177170
15. K. Smolak, K. Siła-Nowicka, J.C. Delvenne, M. Wierzbiński, W. Rohm, Sci. Rep. **11**(1), 15177 (2021)
16. Z. Chen, S. Kelty, A.G. Evsukoff, B.F. Welles, J. Bagrow, R. Menezes, G. Ghoshal, Nat. Commun. **13**(1), 1922 (2022)
17. C. Zhang, K. Zhao, M. Chen, IEEE Trans. Knowl. Data Eng. **35**(5), 4514 (2022)
18. C. Castellano, S. Fortunato, V. Loreto, Rev. Mod. Phys. **81**(2), 591 (2009)
19. D. Brockmann, L. Hufnagel, T. Geisel, Nature **439**(7075), 462 (2006)
20. M.C. Gonzalez, C.A. Hidalgo, A.L. Barabasi, Nature **453**(7196), 779 (2008)
21. V. Colizza, A. Barrat, M. Barthelemy, A.J. Valleron, A. Vespignani, PLoS Med. **4**(1), e13 (2007)
22. L. Hufnagel, D. Brockmann, T. Geisel, Proc. Natl. Acad. Sci. **101**(42), 15124 (2004)

23. S. Ali Alhosseini, R. Bin Tareaf, P. Najafi, C. Meinel, in *Companion Proceedings of the 2019 World Wide Web Conference* (2019), pp. 148–153
24. N. Pantelaios, N. Nikiforakis, A. Kapravelos, in *Proceedings of the 2020 ACM SIGSAC Conference on Computer and Communications Security* (2020), pp. 477–491
25. X. He, Q. Gong, Y. Chen, Y. Zhang, X. Wang, X. Fu, IEEE Trans. Dependable Secure Comput. **18**(5), 2193 (2021)
26. T. Phan-Minh, E.C. Grigore, F.A. Boulton, O. Beijbom, E.M. Wolff, in *Proceedings of the IEEE/CVF Conference on Computer Vision and Pattern Recognition* (2020), pp. 14074–14083
27. J. Ngiam, V. Vasudevan, B. Caine, Z. Zhang, H.T.L. Chiang, J. Ling, R. Roelofs, A. Bewley, C. Liu, A. Venugopal, et al., in *International Conference on Learning Representations* (2021)
28. S. Mozaffari, O.Y. Al-Jarrah, M. Dianati, P. Jennings, A. Mouzakitis, IEEE Trans. Intell. Transp. Syst. **23**(1), 33 (2020)
29. S. Feng, X. Yan, H. Sun, Y. Feng, H.X. Liu, Nat. Commun. **12**(1), 748 (2021)
30. L. Chen, K. Lu, A. Rajeswaran, K. Lee, A. Grover, M. Laskin, P. Abbeel, A. Srinivas, I. Mordatch, Adv. Neural Inf. Proces. Syst. **34**, 15084 (2021)
31. W. Schwarting, A. Pierson, J. Alonso-Mora, S. Karaman, D. Rus, Proc. Natl. Acad. Sci. **116**(50), 24972 (2019)
32. A. Nestor, D.C. Plaut, M. Behrmann, Proc. Natl. Acad. Sci. **113**(2), 416 (2016)
33. A.I. Hsu, E.A. Yttri, Nat. Commun. **12**(1), 5188 (2021)
34. Y. Yan, J.M. Goodman, D.D. Moore, S.A. Solla, S.J. Bensmaia, Nat. Commun. **11**(1), 3564 (2020)
35. J. Shurman, et al., *Calculus and Analysis in Euclidean Space* (Springer, Berlin, 2016)
36. P. Aluffi, *Algebra: Chapter 0*, vol. 104 (American Mathematical Society, New York, 2021)
37. P. Szekeres, *A Course in Modern Mathematical Physics: Groups, Hilbert Space and Differential Geometry* (Cambridge University Press, Cambridge, 2004)
38. C.W. Lynn, D.S. Bassett, Proc. Natl. Acad. Sci. **117**(47), 29407 (2020)
39. S.W. Li, L.C. Xu, C. Zhang, S.Q. Zhang, X. Hong, Nat. Commun. **14**(1), 3569 (2023)
40. M. Chen, H.Y. Ko, R.C. Remsing, M.F. Calegari Andrade, B. Santra, Z. Sun, A. Selloni, R. Car, M.L. Klein, J.P. Perdew, et al., Proc. Natl. Acad. Sci. **114**(41), 10846 (2017)
41. A.P. Bartók, S. De, C. Poelking, N. Bernstein, J.R. Kermode, G. Csányi, M. Ceriotti, Sci. Adv. **3**(12), e1701816 (2017)
42. K. Hansen, F. Biegler, R. Ramakrishnan, W. Pronobis, O.A. Von Lilienfeld, K.R. Muller, A. Tkatchenko, J. Phys. Chem. Lett. **6**(12), 2326 (2015)
43. J.B. Watson, Psychol. Rev. **20**(2), 158 (1913)
44. J.B. Watson, R. Rayner, J. Exp. Psychol. **3**(1), 1 (1920)
45. A. Newel, H.A. Simon, Commun. ACM **19**(3), 113 (1976)
46. C.A. Tisdell, *Concepts of Rationality in Economics* (Institute of Social and Economic Research and Department of Economics ..., 1975)
47. D. Kai-Ineman, A. Tversky, et al., Econometrica **47**(2), 363 (1979)
48. H.A. Simon, Q. J. Econ. **69**(1), 99–118 (1955)
49. R. Thaler, Mark. Sci. **4**(3), 199 (1985)
50. R.J. Shiller, J. Econ. Perspect. **17**(1), 83 (2003)
51. M.H. Bazerman, D.A. Moore, *Judgment in Managerial Decision Making* (Wiley, New York, 2012)
52. R. Thaler, C. Sunstein, in *Amsterdam Law Forum; HeinOnline: Online* (HeinOnline, New York, 2008), p. 89
53. C.F. Camerer, G. Loewenstein, D. Prelec, J. Econ. Lit. **43**(1), 9 (2005)
54. D.J. Benjamin, D. Cesarini, M.J. van der Loos, C.T. Dawes, P.D. Koellinger, P.K. Magnusson, C.F. Chabris, D. Conley, D. Laibson, M. Johannesson, et al., Proc. Natl. Acad. Sci. **109**(21), 8026 (2012)
55. D. Lazer, A. Pentland, L. Adamic, S. Aral, A.L. Barabasi, D. Brewer, N. Christakis, N. Contractor, J. Fowler, M. Gutmann, et al., Science **323**(5915), 721 (2009)
56. C. Song, Z. Qu, N. Blumm, A.L. Barabási, Science **327**(5968), 1018 (2010)

57. C. Stachl, Q. Au, R. Schoedel, S.D. Gosling, G.M. Harari, D. Buschek, S.T. Völkel, T. Schuwerk, M. Oldemeier, T. Ullmann, et al., Proc. Natl. Acad. Sci. **117**(30), 17680 (2020)
58. Y. LeCun, Y. Bengio, G. Hinton, Nature **521**(7553), 436 (2015)
59. A. Newell, P.S. Rosenbloom, J.E. Laird (1989) in *The Foundations of Cognitive Science* (The MIT Press), pp. 97–132
60. N.J. Nilsson, Artif. Intell. **47**(1–3), 31 (1991)
61. T. Winograd, Project MAC. AD0721399. Tech. rep. (1971)
62. T. Winograd, ACM Sigplan Notices **10**(1), 13 (1973)
63. Y. Bengio, A. Courville, P. Vincent, IEEE Trans. Pattern Anal. Mach. Intell. **35**(8), 1798 (2013)
64. G. Wang, X. Zhang, S. Tang, C. Wilson, H. Zheng, B.Y. Zhao, ACM Trans. Web (TWEB) **11**(4), 1 (2017)
65. S.K. Sharma, Sci. Rep. **14**(1), 30830 (2024)
66. Y. Zhu, H. Li, Y. Liao, B. Wang, Z. Guan, H. Liu, D. Cai, in *IJCAI*, vol. 17 (Melbourne, VIC, 2017), pp. 3602–3608
67. J. Feng, L.T. Yang, B. Ren, D. Zou, M. Dong, S. Zhang, IEEE Trans. Comput. **73**(3), 683 (2023)
68. S.J. Clipman, S.H. Mehta, S. Mohapatra, A.K. Srikrishnan, K.J. Zook, P. Duggal, S. Saravanan, P. Nandagopal, M.S. Kumar, G.M. Lucas, et al., Sci. Adv. **8**(42), eabf0158 (2022)
69. Y. Huang, C. Du, Z. Xue, X. Chen, H. Zhao, L. Huang, Adv. Neural Inf. Proces. Syst. **34**, 10944 (2021)
70. J. Wang, Z. Zeng, Y. Wang, Y. Wang, X. Lu, T. Li, J. Yuan, R. Zhang, H.T. Zheng, S.T. Xia, in *Proceedings of the 31st ACM International Conference on Multimedia* (2023), pp. 6548–6557
71. I. Bae, J.H. Park, H.G. Jeon, in *European Conference on Computer Vision* (Springer, Berlin, 2022), pp. 270–289
72. X. Jia, K. Han, Y. Zhu, B. Green, in *Proceedings of the IEEE/CVF International Conference on Computer Vision* (2021), pp. 610–619
73. G. Schwalbe, B. Finzel, Data Min. Knowl. Disc. **38**(5), 3043 (2024)
74. Y. Alufaisan, L.R. Marusich, J.Z. Bakdash, Y. Zhou, M. Kantarcioglu, in *Proceedings of the AAAI Conference on Artificial Intelligence*, vol. 35 (2021), pp. 6618–6626
75. Y. Yang, C. Huang, L. Xia, Y. Liang, Y. Yu, C. Li, in *Proceedings of the 28th ACM SIGKDD Conference on Knowledge Discovery and Data Mining* (2022), pp. 2263–2274
76. C. Xu, M. Li, Z. Ni, Y. Zhang, S. Chen, in *Proceedings of the IEEE/CVF Conference on Computer Vision and Pattern Recognition* (2022), pp. 6498–6507
77. C. Wang, H. Zhu, IEEE Trans. Serv. Comput. **18**, 673–689 (2025)
78. F. Liu, Y. Wen, D. Zhang, X. Jiang, X. Xing, D. Meng, in *Proceedings of the CCS 2019, London, UK, November 11–15, 2019*, pp. 1777–1794
79. J. Wang, R. Wen, C. Wu, Y. Huang, J. Xion, in *Proceedings of the WWW 2019, San Francisco, CA, USA, May 13–17, 2019*, pp. 310–316
80. P. Veličković, G. Cucurull, A. Casanova, A. Romero, P. Lio, Y. Bengio, arXiv preprint arXiv:1710.10903 (2017)
81. B. Yu, H. Yin, Z. Zhu, arXiv preprint arXiv:1709.04875 (2017)
82. C. Wang, H. Zhu, IEEE Trans. Inf. Forensics Secur. **17**, 2703 (2022)
83. C. Wang, H. Zhu, Y. Lin, C. Jiang, arXiv preprint arXiv:2312.01652 (2023)
84. R.F. Bader, J. Phys. Chem. A **114**(28), 7431 (2010)

Chapter 2
Enhancing Behavior Expression Through Behavioral Structure

Abstract The behavioral space is close to infinity, but it can be limited by decomposing behavior into a combination of finite behavioral attributes (just like atoms in molecular structure). This work introduces a general solution, the behavioral molecular structure (BMS) model that characterizes behaviors at the atomic level by transforming the behavioral expression into different combinations of behavioral attributes. A behavioral attribute (like a single atom) is influenced by other behaviors, and perturbations to the behavioral molecular structure can lead to huge differences between behaviors. Correspondingly, this work proposes a GNN-based approach for embedding behavioral molecular structures to obtain highly expressive attribute representations. To demonstrate the effectiveness of behavioral structure, the rationality of behavior molecular structure is examined through theoretical analysis of its expressive power. Furthermore, empirical validation is conducted by deploying this approach in tasks involving fundamental behavioral functions, including detection, prediction, and generation.

2.1 Introduction to Behavioral Molecular Structure

In recent years, the widespread use of graph data in domains such as transportation, biochemical molecular research, and network security has spurred a surge of interest in graph neural networks (GNNs) [1]. Leveraging their message-passing mechanisms, GNNs aggregate information from nodes and their neighbors to effectively capture the complex topological structures of graphs. This capability has enabled GNNs to achieve remarkable performance in various tasks, including node classification, link prediction, and node clustering [2–4].

This part first introduces a general GNN-based solution, the behavioral molecular structure (BMS) model. BMS characterizes behaviors at the atomic level, with the behavioral attributes that make up the behavior being likened to atoms, where the interrelationships between attributes are studied and concretized using graphs. Given a candidate space consisting of potential behavioral attributes, the behavioral molecular structure model can reconstruct behavior as a graph consisting of behavioral attributes and their interactions, based on the interplay between behavioral

C. Wang, *Computational Structural Behavior*,
https://doi.org/10.1007/978-981-95-1013-9_2

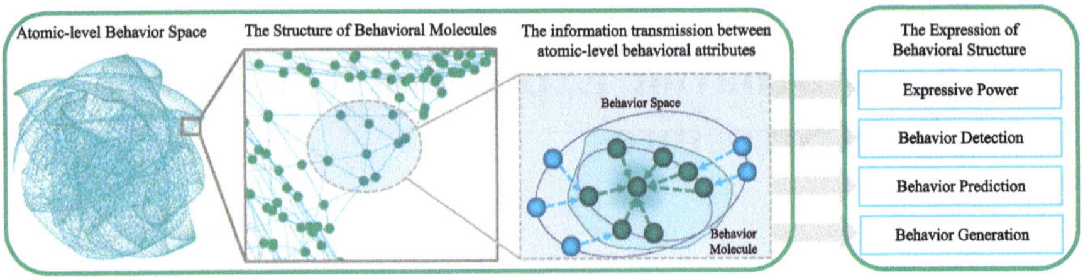

Fig. 2.1 The framework of BMS

attributes (as shown in Fig. 2.1). BMS takes as input a complete behavioral attribute space, specified by all potential behavioral attributes and molecular structures defined on each behavior. The embeddings based on atomic-level behavioral attributes within the behavior molecule and in the contextual environment are repeatedly updated. This process generates a set of embeddings that encode each behavioral attribute. Then, graph pooling is performed on each behavior molecule, and the pooled embeddings are input into additional downstream neural network layers. Finally, we perform downstream tasks (such as detection, prediction, and generation) based on the pooled results obtained from the behavioral molecular structure. By allocating and learning the structure and features of the graph, we introduce an interface for further processing of behavioral structure within existing neural network architectures.

In this part, we make the following contributions:

- To the best of our knowledge, our work is the first work that tangibly captures the interplay between behavioral attributes within behavioral structure. It provides a novel and efficient perspective in utilizing fine-grained attribute information in behavioral data, laying the groundwork for the establishment of behavior models with enhanced expressive power.
- We develop a model called behavioral molecular structure that concretizes the relationships between behavioral attributes as the graph structure and leverages graph learning algorithms to learn representations of behavioral attributes. In this model, all behavior information is shared in the behavioral attribute space, providing high expressive power to support downstream tasks.
- We conduct comprehensive experiments using diverse real-world behavioral datasets, including a crime record category dataset, a user online interaction dataset, and a fraud transaction dataset, to evaluate the performance of our approach on three distinct downstream tasks. The results demonstrate that, compared to traditional behavioral representations, the utilization of behavioral structure leads to enhanced expression of behavior.

2.2 Building Behavioral Molecular Structure

BMS model aims to obtain expressive attribute representations by concretizing fine-grained structures on behavioral attributes. It provides downstream tasks with rich contextual information and can be widely applied in tasks such as behavior detection and prediction. The BMS model consists of four key steps (as illustrated in Fig. 2.2): Firstly, semantic information of behavioral attributes is obtained through a language model. This semantic information can be derived from publicly available knowledge or rely on task-specific domain expertise. Based on the semantics of behavioral attributes, we employ a graph structure to capture the interrelationships between attributes and formalize them as a molecule graph. Subsequently, advanced graph neural network algorithms are utilized to learn representations of the molecule graph, yielding representations for each behavioral attribute. Finally, these representations are applied to specific downstream tasks.

2.2.1 *Definition of Behavioral Space*

Given a set of behavioral data $\mathscr{B}$, the BMS model first needs to map behavioral attributes to a fine-grained behavioral space. Let us define a behavior $b \in \mathscr{B}$

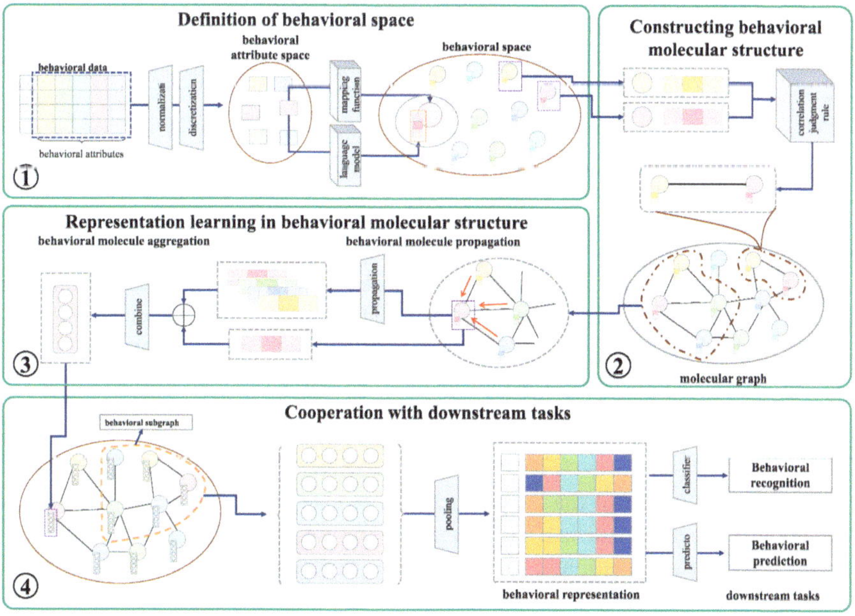

Fig. 2.2 Schematic view of the behavioral molecular structure architecture

as $b = \{a_1, a_2, \ldots, a_m\}$, where m is the number of behavioral attributes in the behavior and $a_i \in \mathscr{A}$ represents the ith behavioral attribute. To establish better connections between behavioral attributes, we map the behavioral attributes to a discrete behavioral attribute space $\mathbb{R}^{|\mathscr{N}|}$, where $\mathscr{N}$ is the set of all discrete behavioral attributes and $|\cdot|$ denotes the cardinality of a set.

2.2.1.1 Behavioral Attribute Mapping

We aim to obtain a behavioral attribute mapping $\phi : \mathscr{A} \rightarrow \mathbb{R}^{|\mathscr{N}|}$. Since behavioral data often originates from diverse sources and has varying specifications, we normalize it to a unified scale and discretize each behavioral attribute into a node in the space $\mathbb{R}^{|\mathscr{N}|}$. This node is denoted as $n \in \mathscr{N}$.

Normalization Attribute normalization is a crucial preprocessing step in the processing of behavioral data. It ensures the compatibility of behavioral attributes across different scales and prevents certain attributes from dominating the subsequent model optimization process due to their large value scales. This normalization process is vital for achieving better model performance and convergence in the subsequent behavioral representation learning process. Additionally, normalization helps to reduce computational costs and improve the generalizability of the model across different behavioral datasets.

For behavioral attributes that exhibit linear relationships, we use the min-max scaling method to transform the attributes into a common scale ranging from 0 to 1. Given a set of attributes of category t, we calculate their regularized attribute values a^t_{norm},

$$a^t_{norm} = \frac{a^t - a^t_{\min}}{a^t_{\max} - a^t_{\min}}, \tag{2.1}$$

where a^t is the original attribute value and $a^t_{\min}$ and $a^t_{\max}$ are the minimum and maximum values of the set of attributes of category t, respectively.

In addition to min-max scaling, we also employ z-score normalization, also known as standardization, which transforms the attributes to have a mean of 0 and a standard deviation of 1. The formula for z-score normalization is given by

$$a^t_{norm} = \frac{a^t - \mu^t}{\sigma^t}, \tag{2.2}$$

where μ^t is the mean of the set of attributes of category t and σ^t is the standard deviation of the set of attributes of category t. z-score normalization is particularly useful when the distribution of the attributes is expected to be normal or Gaussian. The regularization using behavioral attributes is mostly a min-max scaling method since most of the attributes of the dataset used in this work conform to a linear distribution.

Discretization For continuous behavioral attributes in behavioral data (such as temperature and geographical coordinates), each pair of attribute values in different behavioral samples can potentially be distinct. If we simply map them to nodes in the behavioral space using a single function, the resulting number of nodes would be excessively large, which is detrimental for information mining. Therefore, for continuous behavioral attributes, we need to perform discretization. Discretization of continuous variables can be done using various methods.

In this work, we use equal-width discretization with interval

$$width^t = \frac{a^t_{\max} - a^t_{\min}}{num_{bins}}, \tag{2.3}$$

where num_{bins} is the number of manually defined intervals in advance.

During the process of mapping behavioral attributes to nodes in the behavioral space, different types of behavioral attributes require different mapping methods. For instance, the months of the year from January to December can correspond to 12 distinct attribute values in the case of a month attribute. On the other hand, a price attribute can be grouped into bins such as 0–10, 10–20, and 20 or more, representing three different nodes in the behavioral space. By iterating through all the behavioral data, we incorporate a relationship $a \rightarrow n$ in the mapping ϕ for each behavioral attribute a, i.e., $n = \phi(a)$,where n corresponds to a category of attribute values or an interval after binning.

2.2.1.2 Behavioral Semantic Mapping

To extract relevant information from behavioral attributes with semantics, we introduce a behavioral semantic mapping function $\varphi : \mathbb{R}^{|\mathcal{N}|} \rightarrow \mathbb{R}^{|\mathcal{N}| \times dim_s}$, where dim_s represents the dimensionality of the semantic representation. In this work, we extract the semantic information corresponding to behavioral attributes from publicly available background knowledge. Specifically, we employ advanced language models such as BERT to transform attribute nodes n in the behavioral data that contain semantic information into semantic vectors. Language models used for natural language processing are trained on vast amounts of publicly available data and learn the semantic information of behavioral attributes in general contexts, representing this information as dense vectors. BERT, as a deep learning-based language model, generates high-quality semantic representations by considering contextual information and undergoing extensive pre-training and fine-tuning. By leveraging the BERT model, we obtain the semantic vector representation $\varphi(n)$ for each node n. This transformation enhances our understanding of semantic attributes and provides more accurate and comprehensive semantic information for subsequent analysis and mining.

2.2.2 Constructing Behavioral Molecular Structure

After we map the behavioral attributes to the nodes of the behavioral space, we need to construct the behavioral molecular structure in the behavioral space. The main purpose of the behavioral molecular structure is to reflect the attribute associations within the behavior. Specifically, we found that for behavior the various attributes within it are not independent and unrelated to each other. There exists a certain amount of co-occurrence relationship among them. We need to reflect this kind of association in the behavior space. Graph, as a data structure that can naturally express the relationships between nodes, is widely used in various fields. Graph can help us understand and analyze the relationships and networks in complex systems, as well as mine the relationships and information between nodes. Based on this, we establish the connections between nodes in the behavior space to build the molecule graph.

Let the molecule graph in the behavior space be denoted as graph $\mathscr{G}(\mathscr{N}, \mathscr{E}, \mathscr{R})$, where $\mathscr{N}$ denotes the set of nodes (each $n_i \in \mathscr{N}$ represents a node obtained by mapping the behavioral attribute space), $\mathscr{E}$ represents the set of edges (each $e_{ij} = (n_i, n_j, t_i t_j) \in \mathscr{E}$ denotes that node n_i is connected to node n_j and that the type of the connection is $t_i t_j \in \mathscr{R}$), t_i denotes the type of behavioral attribute a_i^t corresponding to n_i, and $\mathscr{R}$ represents the set of types of connectivity relations between nodes. Our goal is to determine whether there is a connection between any two behavioral attribute nodes, i.e., whether there exists an edge e_{ij} between the node pair (n_i, n_j), where $n_i, n_j \in \mathscr{N}$. In other words, the process of constructing our behavioral molecular structure is the process of populating the corresponding adjacency matrix $\mathbf{D}$ of the molecule graph $\mathscr{G}$.

To construct an adjacency matrix for a graph, there are two methods available. The first method, referred to as data association, determines the existence of edges between two behavioral attribute nodes based on the distribution correlation of the attribute vectors. Specifically, we set a threshold δ to determine the similarity between two vectors representing behavioral attribute nodes. If the similarity between these vectors exceeds the threshold δ, an edge is created between the corresponding nodes. Consequently, the adjacency matrix for the behavior molecular structure can be formulated accordingly,

$$\mathbf{D}_{ij} = \begin{cases} 1, & \text{if } ||\varphi(n_i), \varphi(n_j)||_{sim} > \delta \\ 0, & \text{otherwise,} \end{cases} \tag{2.4}$$

where $\varphi(n_i)$ and $\varphi(n_j)$ denote the vectors corresponding to nodes n_i and n_j and $||\cdot, \cdot||_{sim}$ denotes this similarity between the two vectors, which can be euclidean distance, cosine similarity, etc.

The second method is based on the real-life empirical knowledge meta-rules to determine whether there is a correlation between two behavioral attribute nodes. This method relies on the domain experts' empirical knowledge. Specifically, through the experts' experience, we can determine whether there is a correlation between two behavioral attributes and use an adjacency matrix to represent this correlation. In this case, the construction of the adjacency matrix may rely on specific rules and judgment standards, as shown below:

$$\mathbf{D}_{ij} = \begin{cases} 1, & \text{if nodes } n_i \text{ and } n_j \text{ are related in the meta-rules} \\ 0. & \text{otherwise} \end{cases} \tag{2.5}$$

In this work, we opt for using the second method of building the behavioral molecular structure due to the strong real-world characteristics present in the utilized datasets. The use of empirical knowledge and rules allows for a more accurate and reliable determination of the relationship between behavioral attribute nodes. This method harnesses real-world experience and insights to establish meaningful connections, thus enhancing the robustness and applicability of the constructed graph. Additionally, it minimizes the potential biases or limitations that could arise from solely relying on data associations and correlation in the determination of edge existence. Therefore, by leveraging empirical knowledge, we can effectively capture the nuanced and contextual relationships between behavioral attributes, which may not be fully captured by purely statistical methods.

2.2.3 *Representation Learning in Behavioral Molecular Structure*

After obtaining the behavioral molecular structure consisting of behavioral attribute nodes in the behavioral space, we perform representation learning on the structure to obtain the behavioral representation. Specifically, we use the graph learning module to perform message-passing and updating among behavioral attribute nodes.

There are direct and indirect links between certain behavioral attributes. In our construction of the behavioral molecular structure, each attribute is influenced by the attributes of its surrounding neighbors. Our goal in this part is to obtain highly expressive behavioral representations. Specifically, we aim for each attribute to learn semantic and structural information from its surrounding neighboring attributes. Existing methods of graph neural network representation learning have been successfully applied in tasks such as node classification and link prediction, mainly through message-passing and updating mechanisms. Therefore, our behavioral attribute representation learning process is also divided into two sub-modules. The first one is behavioral molecule propagation, which propagates information between the behavioral attribute and its surrounding attributes, thus allowing for attribute-level information intermingling,

$$\mathbf{p}_{\mathbf{v}}^{(k+1)} = \mathrm{ACTF}\left(\mathrm{PROP}\left(\left\{\left(\mathbf{W}_{\mathbf{0}}^{(k)}\mathbf{h}_{\mathbf{v}}^{(k)} \oplus \mathbf{W}_{\mathbf{r}}^{(k)}\mathbf{h}_{\mathbf{u}}^{(k)}\right) | u \in \mathscr{S}_r(v), r \in \mathscr{R}\right\}\right)\right), \tag{2.6}$$

where $\mathbf{p_v}^{(k+1)}$ denotes the information that attribute node v draws from the surrounding attributes at the $(k+1)$th propagation; $\mathscr{S}_r(v)$ denotes the neighbors of node v under a behavioral attribute relation $r \in \mathscr{R}$; $\mathscr{R}$ denotes the set of behavioral attribute relations; ACTF(·) represents the activation function; PROP(·) represents a propagation operation (in this work we use sum), $\mathbf{W_r}^{(k)}$ is the weight of the neighboring node u that is connected to node v through a behavioral attribute relation, and the weight of the target node v is $\mathbf{W_0}^{(k)}$; and $\oplus$ denotes an arithmetic operation (in this work we use addition).

The second one is behavioral molecule aggregation, which updates the intermingled information into new information about the attributes for the next round of behavioral molecule propagation,

$$\mathbf{h_v}^{(k+1)} = \mathrm{COMB}\left(\mathbf{h_v}^{(k)}, \mathbf{p_v}^{(k+1)}\right), \tag{2.7}$$

where COMB(·) denotes the behavioral molecule aggregation operation to update the representation of the behavioral attribute node v.

2.2.4 *Cooperation with Downstream Tasks*

2.2.4.1 Extracting Behavioral Representations

In the constructed behavioral space, the behavioral molecular structure comprises behavioral attribute nodes and edges. Each individual behavior corresponds to a subgraph within this structure, consisting of multiple behavioral attribute nodes. Graph pooling techniques have proven to be highly effective in graph-level learning tasks like graph classification. Hence, we employ a behavioral subgraph pooling module to extract behavioral representations,

$$\mathbf{h_{bh}} = \mathrm{POOL}(\{\mathbf{h_{v_i}} | v_i \in V_{bh}\}), \tag{2.8}$$

where $\mathbf{h_{bh}}$ denotes the behavioral representation, POOL(·) denotes a pooling operation (specifically, mean pooling is employed in this work), $\mathbf{h_{v_i}}$ symbolizes the representation of node v_i within the behavioral subgraph, and V_{bh} signifies the set of nodes in the behavioral subgraph.

2.2.4.2 Optimization Approaches for Different Types of Tasks

Different optimization methods are required when applying behavioral representations to different types of tasks. In particular, supervised learning methods can be employed to optimize behavioral representations for tasks like labeled behavior recognition. By comparing predicted behavioral labels with actual labels, the loss function can be calculated. Optimization algorithms such as gradient descent are then used to adjust model parameters, improving the accuracy and consistency of prediction results. For unlabeled downstream tasks, the model is trained by generating labels from the behavioral data itself to optimize the behavioral representation. We need to employ a classifier whose input is the behavioral representation and output is the predicted label. In this work, we use a three-layer MLP as a classifier with the optimization objective of minimizing the cross-entropy loss function, which measures the difference between the predicted results and the true labels.

2.3 Exploring the Utility of Behavioral Molecular Structure

In this section, we aim to answer the following questions by conducting extensive experiments on three representative datasets:

- **RQ1:** Can the uniqueness be effectively captured in behavioral structure during behavior detection?
- **RQ2:** Can the stability be adequately reflected in behavioral structure for behavior prediction?
- **RQ3:** What role does behavioral structure play in behavior generation tasks?

2.3.1 Uniqueness in Behavioral Structure for Behavior Detection

2.3.1.1 Experimental Setup

To demonstrate the effectiveness of behavioral molecular structure in detection tasks, we utilize a publicly available dataset (Crime Data[1]) consisting of crime incidents in Los Angeles since 2020. In this part, we focus on identifying the 10 most common types of criminal incidents. The original data was transcribed from printed crime reports, with the address field containing information up to the nearest hundred blocks to protect privacy.

Dataset The raw dataset consists of approximately 771K crime incidents, with each incident recorded in 28 columns. These columns include the date of the incident, the geographical identifier of the police station associated with the case, the type of crime, the gender and ethnicity of the perpetrator, and the location of the crime, among others. For time-related columns such as "DATE OCC" and "Date Rptd," we split them into three separate fields representing year, month, and day. Additionally, we extract the hour and minute from the incident time.

[1] *Crime Data:* https://catalog.data.gov/dataset/crime-data-from-2020-to-present.

We also introduce a new field called "Date Difference" that represents the time difference between the date of the crime and the date of reporting. The main reason for incorporating the "Date Difference" attribute is to explore potential patterns or common characteristics among different types of crimes regarding the time interval between the occurrence of the crime and its reporting, considering the victims' perception and response time. For instance, in the case of severe violent crimes, victims are more likely to report them immediately, resulting in shorter date differences. Conversely, for nonurgent property crimes, victims may require more time to detect and confirm the losses, leading to longer date differences. To highlight the differences between different behaviors, we remove the fields "Crm Cd" (crime code) and "Crm Cd Desc" (crime code description) and focus on identifying criminal behavior using the "Crm Cd" field as the target for behavior detection. Ultimately, we select 19 fields for behavior detection: "ARE," "Rpt Dist No," "Part 1-2," "Vict Age," "Vict Sex," "Vict Descent," "Premis Cd," "Weapon Used Cd," "Status," "Cross Street," "Month_Rptd," "Day_Rptd," "Month_OCC," "Day_OCC," "Date Difference," "Hour," "Minute," "Year_OCC," and "Year_Rptd." These fields still include key features such as the timing, location, and victim information within a criminal incident.

Implementation To capture the semantic information associated with the values in these fields, we utilize the pre-trained BERT model ("bert-base-uncased") from Hugging Face library[2]. Here, we input the string values and leverage the pre-trained deep learning neural network to obtain their corresponding 768-dimensional semantic vectors.

We treat the fields in the raw dataset as nodes within the behavioral attribute space. For instance, the dataset includes 12 nodes representing the months, 3 nodes representing the gender of the victims (using the values from the dataset, while our approach allows for a more diverse gender setting), and multiple nodes representing the locations and tools involved in the incidents. To capture the relationships between behavioral attributes, we establish a meta-rule (as illustrated in Fig. 2.3). In Fig. 2.3, nodes represent the values of behavioral attributes, and edges indicate the relationships between two behavioral attributes based on our custom-defined meta-rules. For example, if an incident occurs on December 1 in Block A, where a Caucasian female is robbed using bare hands, connections will be established between the corresponding nodes for December, 1, Block A, bare hands, Caucasian, and female. Within the entire behavioral attribute space, the connections between behavioral attributes can be influenced by multiple behaviors. We aggregate these connections, meaning that the edges that appear more frequently in different behaviors will be assigned higher weights based on their occurrence frequency.

We employ a linear layer to convert the 768-dimensional semantic vectors of all behavioral attributes into the 128-dimensional vectors. Subsequently, we leverage an existing graph neural network algorithm known as RGCN (relational

[2] *Hugging Face:* https://github.com/huggingface.

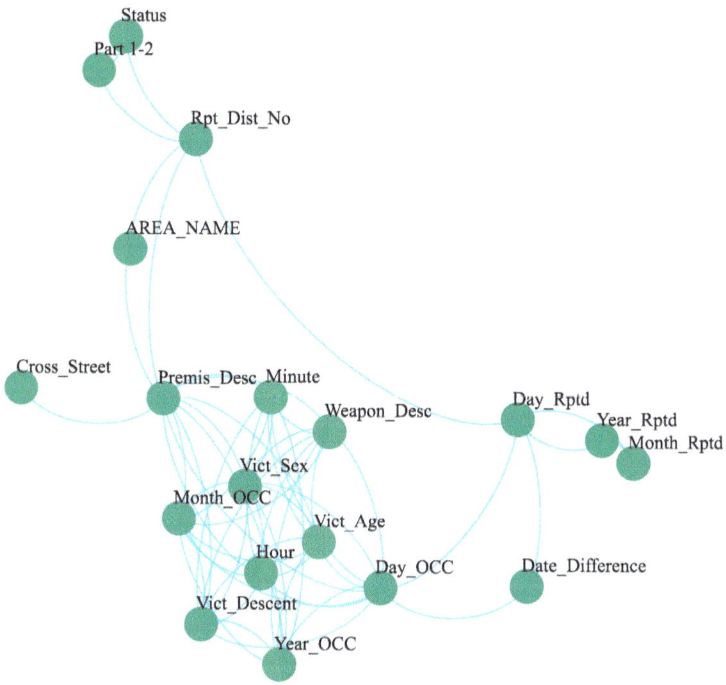

Fig. 2.3 Construction of behavioral molecular structure in Crime Data

graph convolutional network) to facilitate message-passing and aggregation within the structure comprising all known behaviors. Once the message-passing and aggregation process is completed, we obtain the updated node representations that incorporate information from the node itself and its neighbors (including first-order and higher-order neighbors).

In the raw data, each behavior corresponds to a subgraph in the behavioral attribute space, representing the connectivity patterns among several behavioral attributes. We obtain the representation for each behavior by performing average pooling on the node representations associated with the behavior. Subsequently, we utilize a three-layer fully connected neural network to map the behavior to a prediction probability vector. The dimension of this vector depends on the number of behavior categories to be distinguished, where each dimension represents the probability of the behavior vector belonging to a specific behavior category. During the training process, we use cross-entropy as the loss function and employ the Adam optimization algorithm to minimize the loss function. In this case, the initial learning rate is set to 0.001.

2.3.1.2 Result

BMS (referred to as structure) achieves accurate detection by utilizing the structural information of behaviors. It ranks first in five indicators and second in the remaining two indicators, as shown in Fig. 2.4. BMS surpasses typical methods by utilizing behavioral molecular structure on multiple representative classification metrics. The representation-based models generally lag behind BMS but perform better than observation-based models. XGBoost, as another ensemble learning method along with GBDT, shows significant performance improvement and almost matches BMS. This result suggests that although there is a clear expressive power gap between the theoretical upper bounds of structure-based and observation-based methods, the actual performance achieved by models may be far below the ideal performance limit.

In Fig. 2.5, we visualize the trend of changes between the behavioral attribute embeddings learned by BMS and the initial semantic embeddings of behavioral attributes. We can clearly track three different trends. It is like BMS maintains a vector field in the behavioral attribute space, placing each behavioral attribute in a position that better aligns with its context. BMS provides embeddings for each behavioral attribute that possesses contextual information based on the known behavioral data. Deploying downstream behavior detection on these embeddings significantly improves the detection of challenging behavioral data.

In Fig. 2.6a and b, we study models' performance under different numbers of behavioral attributes. We first witness that, in most cases (when the number is greater than 9), all models show performance gains as the number of behavioral attributes increases. This phenomenon can be attributed to the fact that more behavioral attributes provide more information, which helps the models accurately differentiate behaviors. This also aligns with the trend of increasing model expressive power. When the number of behavioral attributes is small (less than 12), BMS performs lower than representation-based methods. Facing sparse structural information within a limited number of behavioral attributes, representation-based approaches

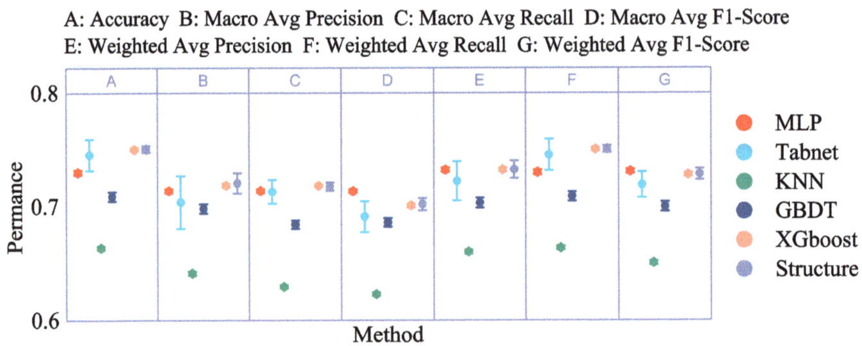

Fig. 2.4 Results in terms of identifying the category of criminal behaviors

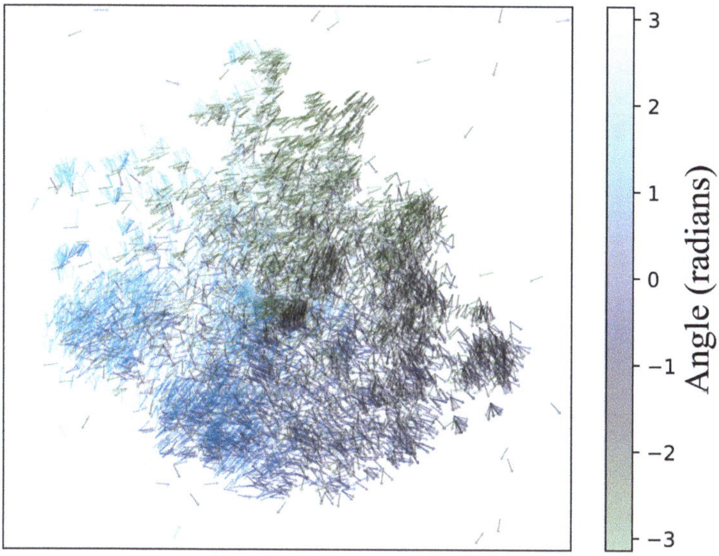

Fig. 2.5 The movement trends between learned vectors and initial semantic vectors

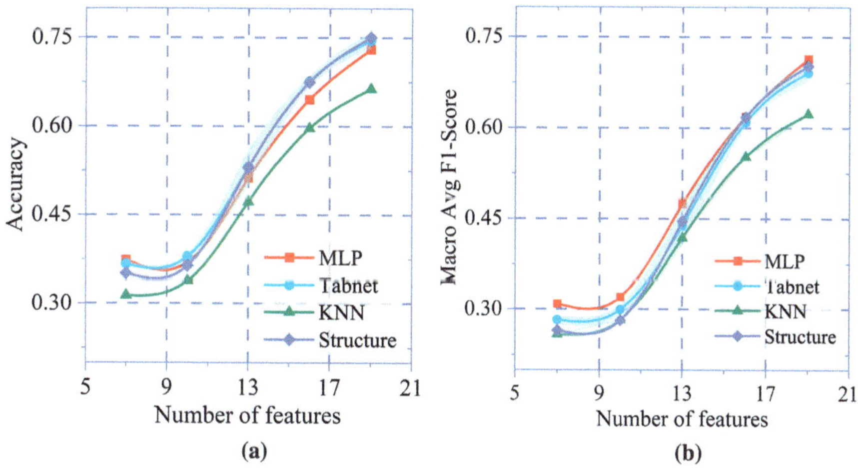

Fig. 2.6 The trend of accuracy and macro average F1-score in behavior detection as the number of behavioral attributes changes. (**a**) Accuracy. (**b**) Macro average F1-score

can extract sufficient information for behavior detection even with rough information mining. However, as the number of behavioral attributes increases, the importance of structural information becomes more pronounced, and representation-based methods gradually lag behind structure-based methods.

Data-driven models may lead to biased outcomes in public domains [5]. Specifically, we investigate whether BMS can capture the distribution of criminal behaviors

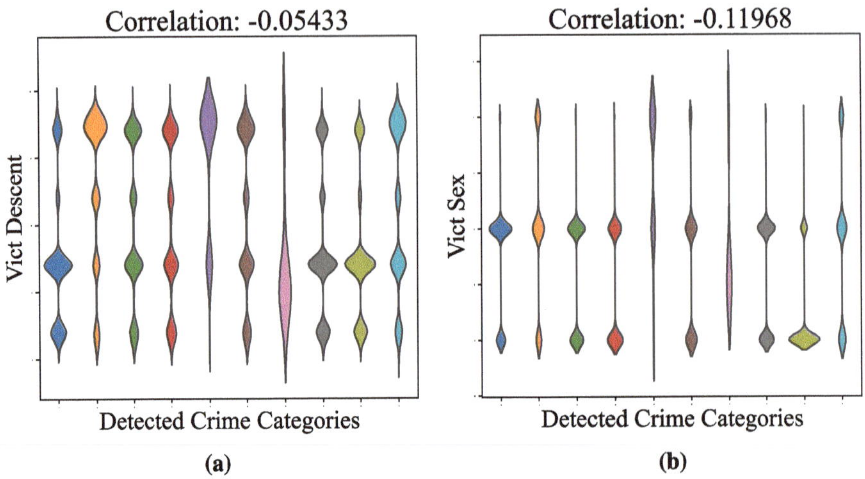

Fig. 2.7 The correlation between the crime behavior categories predicted by BMS and the ethnicity/gender of the victims. (**a**) Ethnicity. (**b**) Gender

fairly. From Fig. 2.7a and b, we find that there is a low correlation between gender and ethnicity with the detected types of criminal behaviors, which is consistent with the original data. Further examining each specific criminal behavior, we note two anomalous phenomena: criminal behavior 1 ("THEFT FROM MOTOR VEHICLE - PETTY ($950 & UNDER)," represented in purple) and criminal behavior 2 ("VEHICLE-STOLEN," represented in pink) are not proportionally captured by the models. To investigate the cause of these anomalies, we analyze the statistical information regarding different ethnicities and genders of victims in the models' identification of criminal types (Fig. 2.8).

Regarding criminal behavior 1, the number of correctly identified instances indicates that both the structure and XGBoost exhibit weak detection capability (the accuracy of only 0.00636 and 0.04495). We speculate that this might be due to models misclassifying such behaviors as other criminal types. The detection of these particular criminal behaviors is often considered to have lower reliability. Regarding criminal behavior 2, the limited number of instances of such behaviors hinders the accurate identification of their criminal types. Similarly, this phenomenon also exists concerning the gender factor of the victims.

2.3.2 *Stability in Behavioral Structure for Behavior Prediction*

2.3.2.1 Experimental Setup

We conduct a study using real user online interaction data to investigate the performance of BMS in behavior prediction. On the knowledge-sharing platform

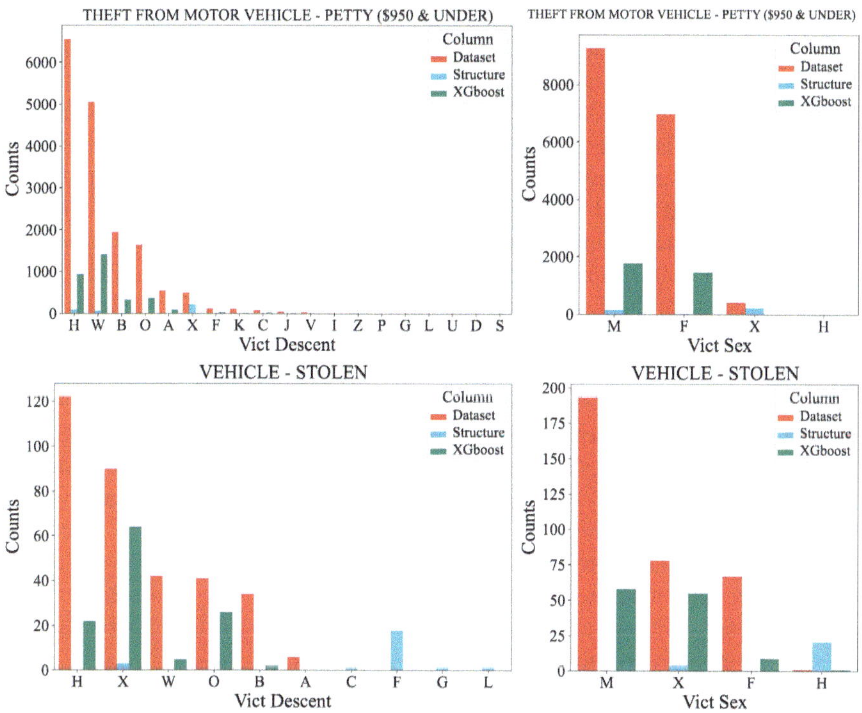

Fig. 2.8 The statistical data regarding the identification of criminal behaviors ("theft from motor vehicle-petty ($950 & under)" and "vehicle stolen") by different models

(Zhihu), users are recommended several answers related to questions they are interested in, and they choose to browse the answers they find interesting. By modeling users' interaction behavior based on their historical interactions, we study the prediction of users' potential choices for their next interaction.

Dataset The ZhihuRec Data[3] is a real-world personalized recommendation interaction dataset used exclusively for scientific research. It was jointly constructed by the information retrieval group of Tsinghua University (THUIR) and Zhihu company (a knowledge-sharing platform). In our work, we focus on studying the performance of behavior substructure in behavior prediction based on the ZhihuRec dataset. The dataset consists of approximately 100 million interaction logs, capturing various users, answers, questions, authors, and topics within the Zhihu question and answer community. We first perform data filtering and aggregation on the original dataset and then deploy the behavioral molecular structure model based on the newly constructed fields.

[3] *ZhihuRec Data:* https://github.com/THUIR/ZhihuRec-Dataset.

Implementation We mainly utilize the "info_user.csv," "info_answer.csv," and "inter_impression.csv" files from the dataset, representing user information, answer information, and user-answer interaction behavioral data, respectively. Based on the "inter_impression.csv" file, we establish the interaction relationship between user IDs and answer IDs to establish the connection between users and answers. Furthermore, in the "info_answer.csv" file, we extract the one-to-one correspondence between answer IDs and answer contexts. Additionally, when an answer appears on the user interface, a timestamp is generated to record this event (corresponding to "impression timestamp" in the "inter_impression.csv" file). Furthermore, when a user accesses an answer, the timestamp of the user's click is recorded in the "inter_impression.csv" file (denoted as "click timestamp"). With this, we can establish the association between a user accessing an answer at a specific moment. Then, we split the "click timestamp" and "impression timestamp" fields by hour. Similarly, for the "register timestamp" field in the "info_user.csv" file, we split it by month. Both answer topics and user topics have the same value range, consisting of topic IDs. Therefore, we consider them as a single field called "topic," and we can establish the association between a user accessing an answer under a specific topic. Taking into account the relevant information for users and answers, we establish associations with the user ID and answer context, respectively. In summary, when a user initiates a click on an answer, its corresponding substructure can be modeled as a graph structure (as shown in Fig. 2.9).

In behavior prediction, we employ the graph neural networks as the learning module for behavior molecular structure. Specifically, we use a two-layer RGCN layer with the activation function of ReLU and a linear layer to obtain the final output embeddings. Based on the distribution of user clicks, we classify users into *four* categories based on clicks of 0–19, 20–49, 50–99, and 100 and above, as labels for node classification when learning output embeddings. In this work, we set the embedding dimension to be 64.

After obtaining the embeddings of the nodes, we extract 64-dimensional vectors for each user ID and answer ID, which serve as features for subsequent tasks. For the subsequent tasks, we utilize the RecBole[4] platform and adopt its sequential recommendation method as our prediction module. Specifically, leveraging the interaction data and features of users and answers, we recommend answers to users using the sequential recommendation method. This entails predicting whether a user will click on a specific answer. If the user indeed clicks on the answer, it is considered a correct prediction. By utilizing the RecBole platform, we use its sequential recommendation method to make predictions based on user-answer interactions and features. This allows us to effectively predict the behavior of users, specifically whether they will click on a given answer, and assess the performance of our model.

[4] *RecBole:* https://recbole.io.

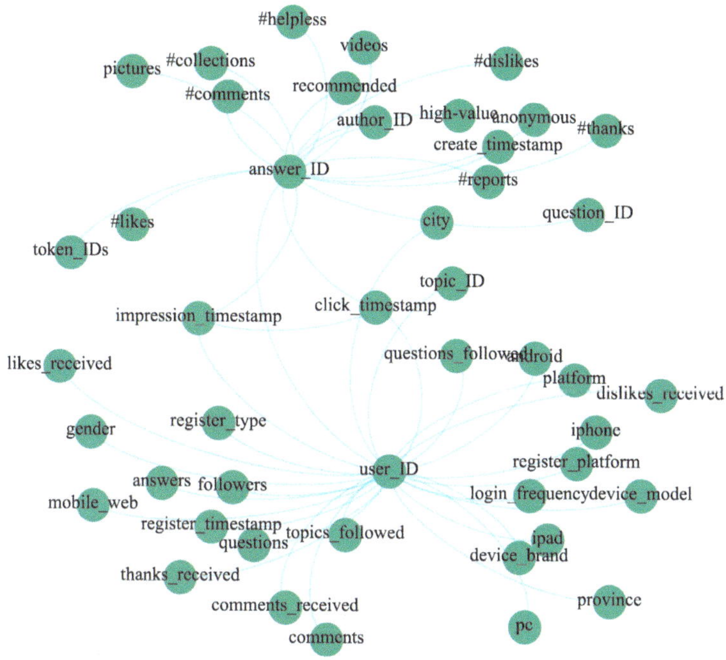

Fig. 2.9 Construction of behavioral molecular structure in ZhihuRec Data

2.3.2.2 Result

The results indicate that BMS exhibits unstable performance in predicting user-interest answers (as shown in Fig. 2.10). BMS achieves unstable performance by utilizing behavioral molecular structure that represent multiple recommendation metrics. Hollow points in the figure represent the performance of baseline methods, while solid points represent the performance of baseline methods after leveraging BMS to learn behavioral structure. For models such as GRU4Rec, STAMP, and FEARec, BMS consistently helps surpass their original performance. Unfortunately, when faced with FPMC, TransRec, and CORE, BMS fails to filter out noise information, leading to the performance decline. The Pop and ItemKNN methods only utilize user-answer interactions for prediction, which prevents BMS from incorporating structural information into these models. Despite using limited information, these two models outperform others in multiple metrics. This phenomenon also confirms that the strength or weakness of model expressive power is not strictly correlated with its performance in specific downstream tasks.

Information entropy, describing the uncertainty of possible behavior occurring in an information source, can be used to capture the predictability of user behavior sequences [6]. We introduce time-independent entropy S, $S = -\sum_{j=1}^{N_i} p_i(j)\log_2 p_i(j)$, to measure the predictability of user behavior based on

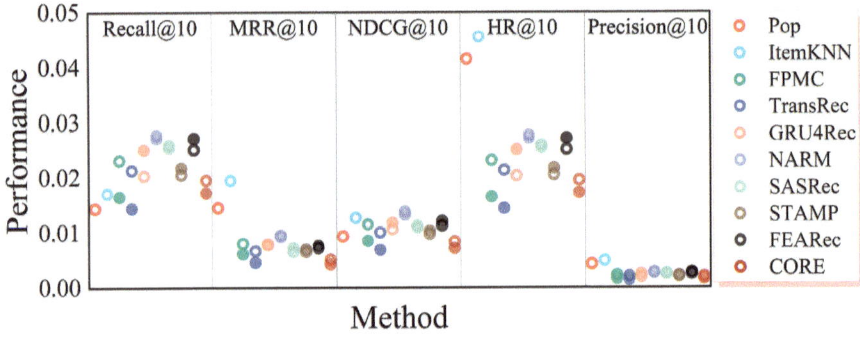

Fig. 2.10 Results in terms of predicting user interaction behaviors

their historical interaction records, where $p_i(j)$ represents the historical probability of user i accessing the answer j. We randomly select 200 users not included in the training data, analyze their interaction behaviors in consecutive time intervals, and calculate the time-independent entropy. The results are shown in Fig. 2.11. We perceive that with the increase in known behavior the predicted time-independent entropy shows a decreasing trend, i.e., the uncertainty of user behaviors decreases, making it easier to predict for users with more historical records. Due to the inherent randomness in user behavior sequences, the time-independent entropy of users increases at the second moment instead of showing the expected decreasing trend. But at the third moment, the time-independent entropy of users quickly decreases and even falls below the average value of the predicted results. The predictability of behavior only indicates the probability of correctly predicting a user's future trajectory with an appropriate prediction algorithm.

2.3.3 Potential Opportunities in Behavioral Structure for Behavior Generation

2.3.3.1 Experimental Setup

We deploy BMS to generate behaviors not included in the training set using limited behavioral data for behavior generation and detecting the effectiveness of its generated behavior. Here, part of the behavioral structure will be kept unseen until BMS completes information passing and model training based on known behavioral structure.

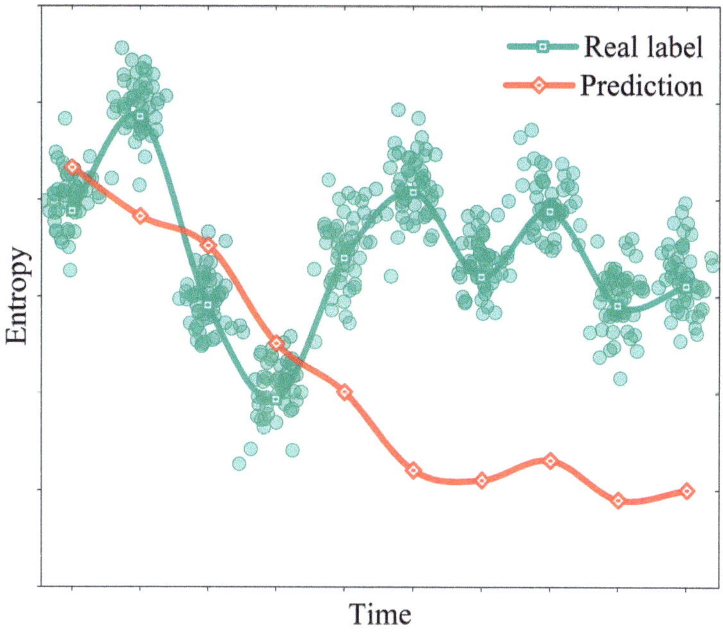

Fig. 2.11 Visualization of predictability of user behavior

Dataset We select a fraud transaction dataset[5] from a bank in South Africa, where fraud behaviors refer to illegal profit from massive transfers from one account to another. The dataset contains 6,362,620 behavioral transactions consisting of 10 fields that involve transfers between bank accounts. We perform the following preprocessing steps on the original data. Since the field step is naturally a discrete integer representing the time period encoding of when the transaction occurred, no further processing is needed. For string-type fields such as type, nameOrig, and nameDest, inspired by natural language processing, we convert the original text data into tagged format. For amount-type fields such as amount, oldBalanceOrg, newbalanceOrig, oldbalanceDest, and newbalanceDest, we label them and then bucket them according to their maximum counting unit in decimals.

Implementation We map each behavior to a subgraph structure consisting of a set of nodes and edges, where different fields in the original behavioral dataset are considered as different types of nodes. Therefore, each behavior can generate a corresponding heterogeneous graph. Based on our understanding of anti-fraud business processes, we define metarules by selecting corresponding fields to represent nodes and selecting specific node pairs to represent edges, thus generating behavioral

[5] *Fraudulent transaction data:* https://www.kaggle.com/datasets/chitwanmanchanda/fraudulent-transactions-data.

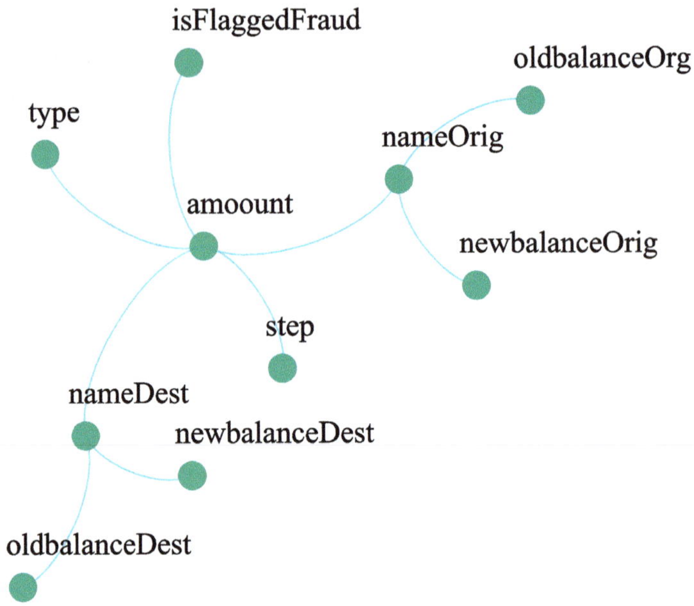

Fig. 2.12 Construction of behavioral molecular structure in fraudulent transaction data

molecular structure containing 11 different types of nodes and 11 different types of edges (as shown in Fig. 2.12)—the steps for behavioral structure generation that we utilized an existing method called GraphVAE.

The fraudulent transaction data is a highly imbalanced dataset, with 8,213 fraudulent transaction records and 6,354,407 normal transaction records. In our fraud prediction task, we select all fraudulent transaction records from the original dataset as well as a dataset of normal transactions ten times the size of the frauds for experimental verification. We consider fraud detection as a binary classification task and use AUC as our main evaluation metric. Additionally, we also count the amount of money involved in the successfully identified fraudulent transactions.

We design two different generative strategies to validate the capabilities of BMS:

- S1: We only introduce the generated structures during training while using real data to validate the BMS's performance of fraud detection.
- S2: We use the original data during training and mix in fraud behaviors based on generated structures during testing to validate the BMS's performance of fraud detection.

2.3.3.2 Result

We adopt a generative-based method to infer the unseen structures in the behavioral attribute space (Fig. 2.13). In Fig. 2.14, we introduce graph structure-related metrics to compare the known structures with those generated by BMS. The small KSI and Orbit indicate that BMS has learned approximate structural information. The KSI and Orbit reflect that the generated behavioral structure reserves the expected structure and has a distribution very close to the original structure. The two other metrics, Novel and Unique, reflect the generalization ability in learning behavioral structure [7]. Large values mean that the generated structure contains structures that do not exist in the original data. The results indicate that BMS can effectively capture previously undiscovered structural information.

We further design two groups of experiments to validate the effectiveness of the generated structures by hiding a portion of the raw behavioral data. One is to use the generated structures for the training process, and the other is to use the generated structures for the testing process. Ideally, BMS should generate potential associations among behavioral attributes based on associations with known structures, and these associations can help downstream tasks (Fig. 2.15). We observe that the structural information generated from BMS's inference from a part of the known behaviors is less than the structural information contained within the data itself and the performance of downstream fraud detection decreases with the increasing proportion of hidden behaviors. In Fig. 2.15a, performance continues to

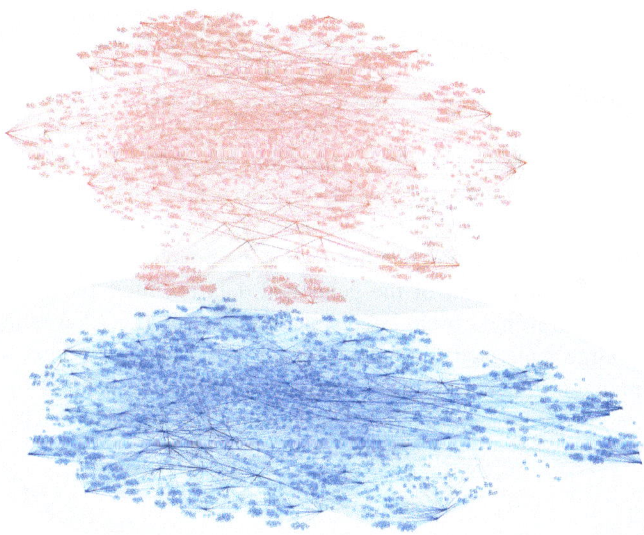

Fig. 2.13 The generated behavioral molecular structure (blue) by BMS with high similarity compared to the determined structure (orange) in the data

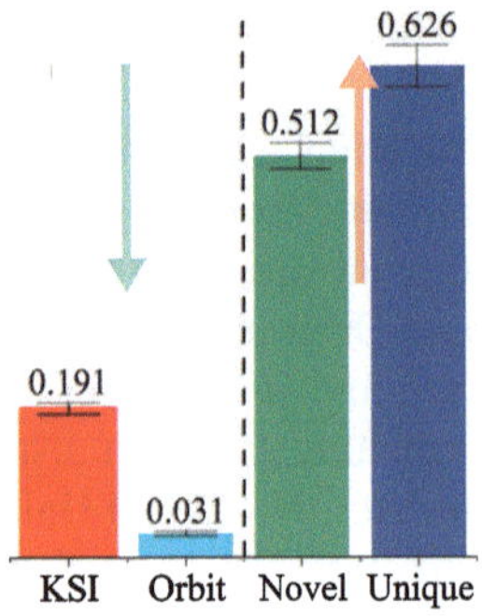

Fig. 2.14 The generated behavioral molecular structures based on limited behavioral data by BMS

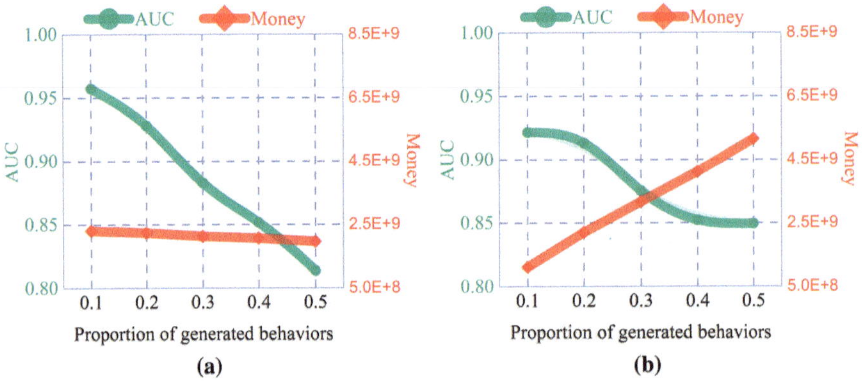

Fig. 2.15 Results in structural evaluation

decline gradually as more generated structures are used for training; the decrease in the amount of money involved in fraudulent transactions is minor. It indicates that BMS's ability to learn from small transactions is limited, while it can effectively identify key large transactions by capturing crucial structures (similar to functional groups crucial to molecular structures). In Fig. 2.15b, as more generated structures are used for testing, we find that the decreasing trend nearly stops when they exceed 40%, meaning that most of the generated data are identifiable. Additionally, we note that the prevented fraud losses increase linearly (i.e., downstream models can identify large generated fraudulent behaviors). These experiments show that for high-amount transactions, BMS learns crucial structural compositions in generated structures, while for small-amount transactions, BMS's structural learning is limited due to their scattered and insignificant features.

2.3.4 Expressive Power of Behavioral Molecular Structure

In our work, we propose a modeling strategy called the behavior molecular structure, which aims to enhance the expressive power of behavior models. We validate the effectiveness of the strategy in terms of expressive power within the context of limited behavior dimensions. As the dimension continues to increase, it will even grow indefinitely, and we anticipate that the performance gains will persist and be more advantageous in higher dimensions. In this context, we try to give a reliable prediction around the growing behavioral dimension.

In deep learning, the expressive power of a model can be estimated by the count of linear regions that are partitioned by the model [8]. Models with more linear regions tend to better capture relationships within the data. We illustrate with an example of a single hidden layer fully connected neural network. In this network, each unit in the hidden layer can operate in two modes: a nonactivated state where it takes a zero value, and an activated state where it takes a nonzero value. The boundary between these two modes is defined by a specific hyperplane. Consequently, the number of linear regions in a single-layer MLP is equal to the number of regions formed by the collection of hyperplanes.

Similarly, we introduce the concept of measuring behavioral expressive power by considering the number of distinct behaviors into which models partition the behavioral space. In this context, a behavior is a set of features (referred to as behavioral attributes) that are used for behavioral description. We denote γ as the number of behaviors that can be expressed in the behavior space Fig. 2.16 illustrates the behavioral expressive power of three representative models, observation-based, representation-based, and structure-based models, where we implement logarithmic operation on the value of expressive power to facilitate display.

Observation-based models exhibit a constant level of behavioral expressive power as they are irrespective of behavioral dimension. Therefore, we can denote this level as $\gamma_O = C$, where C represents a constant value; the corresponding magnitude is $\mathcal{O}(1)$.

Representation-based models provide an exponential level of behavioral expressive power. They can divide the behavioral space into the number of behavioral value combinations. Given behavioral data of dimension m, where the number of values for the behavioral attribute on the ith dimension is denoted as k_i, the number of all possible behaviors γ_R can be calculated as

$$\gamma_R = \prod_1^m k_i \leq \prod_1^m k = k^m, \tag{2.9}$$

where $k := \max(k_1, \cdots, k_m)$. By scaling k_i, the theoretical upper limit of behavioral expressive power for such models is k^m (k is set to 100 in Fig. 2.16); the corresponding magnitude is $\mathcal{O}(k^m)$.

Structure-based models enable the translation of behavioral expression into nodes' connectivity in the graph. Given behavioral data of dimension m, where a behavior corresponds to m behavioral attribute nodes, the lower bound on behavior's

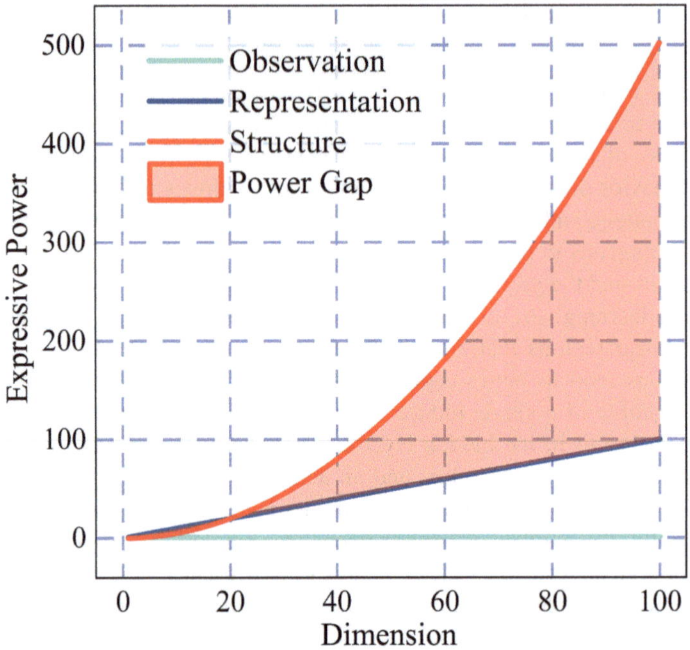

Fig. 2.16 Comparison of expressive power of different behavioral expressions on behavioral dimensions

number is determined by the number of subgraphs, i.e., $\gamma_S \geq k^{m\cdot(m-1)}$; the corresponding magnitude is $\mathcal{O}(k^{(m^2)})$. Here, $m(m-1)$ represents the possible states of connectivity between nodes, such as connected or unconnected, and k represents the number of values that a node can take with $k = 2$ in Fig. 2.16. We can calculate that the difference in expressive power between structure-based models and representation-based models is positively correlated with

$$\gamma_S - \gamma_R \propto \mathcal{O}(k^{(m^2)}) - \mathcal{O}(k^m) = \mathcal{O}(k^{(m^2)} - k^m) = \mathcal{O}(k^{(m^2)}). \tag{2.10}$$

Hence, in most cases ($m \geq 2$), the structure-based models can theoretically surpass ordinary exponential level $\mathcal{O}(k^m)$. It is crucial to emphasize that possessing a high level of expressive power does not guarantee superior performance and should not be considered the sole indicator of high performance.

The notable result is that with the increase of behavioral dimension behavioral structure has a significant advantage compared to the other two ways of expression. We also notice that when the behavioral dimension is less than 20, the expressive power of behavioral structure is slightly lower than that of behavioral representation. This shows that when behaviors only require relatively few features to describe, facing such a simple behavior, behavioral representation can sufficiently describe the behavior.

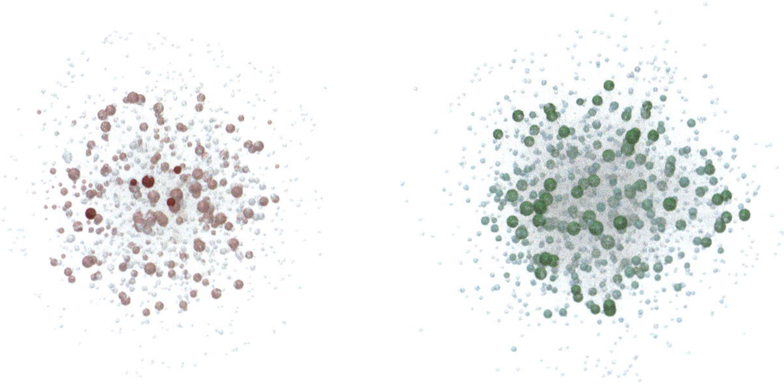

(a) (b)

Fig. 2.17 Demonstration of the expressive power of behavioral molecular structure

Representation-based models can use the similarity between representations to compare the semantic information of two behaviors. We use a similar method to estimate the semantic influence between behaviors in structure-based behavioral research. To illustrate this, we visualize 2,000 criminal behaviors from an open dataset. To ignore the effects of structure selection, we assume that there is the correlation between any two behavioral attributes within each behavior. We study the influence of behavioral attributes on contextual behaviors and their impact on behavioral semantics. In Fig. 2.17a, we use a visualization algorithm to perform visualization of the structure of 2000 behaviors each consisting of 10 behavioral attributes and randomly select one behavior to colorize based on the neighbor of each of its behavioral attributes. The resulting graph provides an example of how behaviors are positioned in behavioral space and how they influence contextual behaviors. We find that a behavior's semantic information can have an impact on almost the entire behavioral attribute space through indirect correlations between behavioral attributes. In Fig. 2.17b, behavioral attribute frequency is associated with these behaviors shown in Fig. 2.17a. The area of a node corresponds to the number of times the corresponding behavioral attribute appears in the behavioral data, and the degree of a node is roughly proportional to its area. Behavioral attributes that appear frequently tend to be located in key positions in the behavioral structure. It shows that frequently occurring behavioral attributes tend to be located in key positions within behavioral attribute communities, with more connections transmitting their information to other behaviors.

2.4 Denoising in Behavior Molecular Structure

In practical scenarios, graph structures are often extracted from complex interacting systems that often contain noise, errors, and missing connections, which significantly constrain the performance and applicability of GNNs [9]. Moreover, the reliance of GNNs on explicit graph structures limits their applicability in broader scenarios.

To address the issue of noise, we propose an innovative approach that models behavioral tabular data as behavior molecule graphs. This method constructs small-scale molecule graphs based on attribute associations, enabling the capture of fine-grained structural information while mitigating the propagation of noise through message-passing mechanisms. By avoiding the noise amplification inherent in large-graph modeling, this approach offers improved robustness. However, the molecule graphs may still contain structural biases introduced by manual definitions. To further optimize graph structural information, this study integrates graph structure learning (GSL), employing parameterized adjacency matrix modeling and contrastive learning strategies to enhance the robustness of graph structures and improve performance in downstream tasks [10].

Within the GSL framework, we propose a novel method called behavior molecular structure denoising (BMSD), which leverages self-supervised contrastive learning to optimize the quality of behavior molecule graphs. Specifically, BMSD maximizes the mutual information between the learned structure and a predefined anchor graph, uncovering latent connections while effectively reducing noise. This approach significantly enhances performance in behavior recognition tasks. Experimental results highlight the effectiveness of this method on real-world datasets, offering a new perspective for structural optimization and deep modeling of tabular data. In summary, our study is the first to integrate structural information and optimization for tabular data, presenting an unsupervised framework that shows great potential in fine-grained graph data modeling and performance improvement in downstream tasks.

Next, this section presents the behavior molecular structure denoising framework based on graph structure learning. The framework employs a self-supervised graph contrastive learning approach to optimize and denoise structural information in behavior molecule graphs by constructing structural contrasts. The method consists of three core modules, each addressing specific tasks: generating behavior molecule graphs, constructing contrastive views, and optimizing graph structures through contrastive learning.

2.4.1 Behavior Molecule Graph Construction Module

In real-world scenarios, the limited number of features and constraints such as privacy protection often results in observed behavioral features failing to capture all dimensions of behavior comprehensively. With the advancement of deep learning, there is an increasing demand to leverage the rich information inherent in data to enable models to effectively learn the complex patterns underlying tasks. To better analyze and understand the implicit information within data, facilitating more efficient learning for various behavior detection and recognition tasks, we adopt a behavior molecule modeling approach. This method constructs independent behavior molecule graphs for each data point based on attribute distributions, incorporating fine-grained structural information to uncover latent patterns in the data.

Let X denote a behavioral dataset consisting of behavioral records, where each behavior is described by multiple attribute features. A behavioral record x is represented as $x_k = \left(x_k^1, x_k^2, \ldots, x_k^i, \ldots, x_k^n\right)$, where x_k^i denotes the ith attribute feature of the kth record. The graph structure is defined as $\mathcal{G} = (A, P)$, where A is the adjacency matrix and $P \in \mathbb{R}^{N \times n}$ is the node feature matrix.

In the behavior molecule graph, nodes represent attribute features of the behavior, and edges represent associations between these attributes. The behavior molecule model construction module transforms tabular behavioral data into an initial behavior molecule graph. The feature matrix $P = [p_1, p_2, p_3, \ldots, p_n]$ encodes the n-dimensional attribute information of the behavior. The connectivity score matrix S quantifies the connectivity of the behavior molecule graph, representing relationships between attributes:

$$\begin{cases} p_i = & x_k^i, \\ s_{ij} = & \frac{N_{ij}}{N_i + N_j}, \end{cases} i, j \in \mathcal{N} \tag{2.11}$$

Based on the connectivity score matrix S, we derive the adjacency matrix $A = \left(n_{ij}\right)_{n \times n}$ for each behavior molecule graph. To constrain the size of each molecule graph, a connectivity score threshold c is introduced. Specifically, if $s_{ij} < c$, then $n_{ij} = 0$; otherwise, $n_{ij} = 1$.

2.4.2 *Behavior Molecule Graph Structure Learning Module*

The graph learner, a key element in graph structure learning, produces a parameterized sketch adjacency matrix $\tilde{\mathbf{S}} \in \mathbb{R}^{n \times n}$. Since our data comprises behavior molecule graphs, where each node corresponds to an attribute from original behavioral tabular data and the graphs are individual small graphs rather than a single large graph, conventional graph learners such as those based on node similarity metrics or graph neural networks are ineffective. To obtain a high-quality sketch adjacency matrix, we utilize the connectivity score matrix S as the parameterized sketch adjacency matrix:

$$s_{ij} = \frac{N_{ij}}{N_i + N_j}, i, j \in \mathcal{N}, \tag{2.12}$$

where $\mathcal{N} = 1, 2, \ldots, n$, n represents the behavioral dimensions and s_{ij} reflects the distribution of attribute pairs across the dataset, serving as an indicator of the association between attributes.

The first step is sparsification. The generated preliminary adjacency matrix $\widetilde{\mathbf{S}}$ is densely populated, reflecting a fully connected graph. This type of matrix is typically impractical for most applications and incurs high computational costs. To address this, we apply k-nearest neighbor-based sparsification to $\widetilde{\mathbf{S}}$. For each node, only the top k edges with the highest connection values are preserved while the weights of all other edges to zero. The sparsification function $q_{sp}(\cdot)$

$$\tilde{\mathbf{S}}_{ij}^{(sp)} = q_{sp}(\tilde{\mathbf{S}}_{ij}) = \begin{cases} \tilde{\mathbf{S}}_{ij}, & \tilde{\mathbf{S}}_{ij} \in top - k(\tilde{\mathbf{S}}_i), \\ 0, & \tilde{\mathbf{S}}_{ij} \notin top - k(\tilde{\mathbf{S}}_i), \end{cases} \tag{2.13}$$

where top-k($\widetilde{\mathbf{S}}_i$) represents the set of the top k values in the row vector $\widetilde{\mathbf{s}}_i$. Since graph learning is computationally intensive, applying sparsification yields a sparser graph structure.

Symmetrization and Activation: Real-world graphs generally exhibit bidirectional connections, making a symmetric adjacency matrix essential. Moreover, as per the definition of adjacency matrices, edge weights must be nonnegative. To fulfill these requirements, the following are the steps for symmetrization and activation,

$$\tilde{\boldsymbol{S}}^{(sym)} = q_{sym}\left(q_{act}(\tilde{\boldsymbol{S}}^{(sp)})\right) = \frac{\sigma_q(\tilde{\boldsymbol{S}}^{(sp)}) + \sigma_q(\tilde{\boldsymbol{S}}^{(sp)})^T}{2}, \tag{2.14}$$

where $\sigma_q(\cdot)$ is the ReLU function in this model.

Normalization: To constrain the edge weights within [0, 1], a final normalization step is applied to $\widetilde{\mathbf{S}}$. Specifically, symmetric normalization is employed,

$$\boldsymbol{S} = q_{norm}(\widetilde{\boldsymbol{S}}^{(sym)}) = (\widetilde{\boldsymbol{D}}^{(sym)})^{-\frac{1}{2}} \widetilde{\boldsymbol{S}}^{(sym)} (\widetilde{\boldsymbol{D}}^{(sym)})^{-\frac{1}{2}} \tag{2.15}$$

where $\widetilde{\mathbf{D}}^{(sym)}$ denotes the degree matrix of $\widetilde{\mathbf{S}}^{(sym)}$.

2.4.3 Structure-Guided Contrastive Learning Module

In the previous module, we obtained an adjacency matrix S. The subsequent task is to implement the structure-guided contrastive learning module. Two graph views are constructed based on the learned structure and the original data. Subsequently, data augmentation is applied to these two views to improve the model's robustness and generalization. Finally, we maximize the MI between two views, optimizing the model's performance across different views, ensuring that the learned representations are more consistent and meaningful.

The contrastive views are directly constructed by combining the learned adjacency matrix S and the feature matrix X, represented as $\mathcal{G}_l = (S, X)$. We initialize the contrastive view as a k-nearest neighbor graph based on the features. The anchor view serves as a teacher in graph structure learning, offering stable and accurate guidance. In the structural denoising task, the anchor view is $G_a = (A_a, X) = (A, X)$.

In the BMSD framework, we employ two simple yet effective data augmentation methods. To perturb node features, a random subset of feature dimensions is selected and set to zero, simulating different feature loss scenarios. This method introduces noise, which helps improve the model's robustness and encourages it to learn more generalized representations. We apply masking vector $m^{(x)} \in 0, 1^d$ to mask the feature vectors,

$$\overline{X} = T_{fm}(X) = [x_1 \odot m^{(x)}, \cdots, x_n \odot m^{(x)}]^{\mathrm{T}}, \tag{2.16}$$

where X is the extended feature matrix.

We also disrupt the graph structure by randomly dropping edges. The adjacency matrix is masked with $M^{(a)} \in 0, 1^{n \times n}$,

$$\overline{A} = T_{ed}(A) = A \odot M^{(a)}, \tag{2.17}$$

where A represents the adjacency matrix.

Both augmentation strategies are simultaneously employed to create enhanced graphs for the contrastive and anchor views,

$$\overline{\mathcal{G}_l} = \left(\mathcal{T}_{ed}(\mathrm{S}), \mathcal{T}_{fm}(\mathrm{X})\right), \overline{\mathcal{G}_a} = \left(\mathcal{T}_{ed}(\mathrm{A}_a), \mathcal{T}_{fm}(\mathrm{X})\right), \tag{2.18}$$

where $\mathcal{G}_l$ and $\mathcal{G}_a$ represent the augmented learner view and anchor view, respectively.

Once the two augmented graph views are generated, a node-level contrastive learning method is employed to maximize their MI. A GNN-based encoder $f_\theta(\cdot)$ is used to extract node-level representations from the augmented graphs $\bar{\mathcal{G}_l}$ and $\bar{\mathcal{G}_a}$,

$$H_l = f_\theta(\overline{\mathcal{G}_l}), H_a = f_\theta(\overline{\mathcal{G}_a}), \tag{2.19}$$

where θ represents the parameters of the encoder $f_\theta(\cdot)$.

After the encoder, we employ a projector $g_\varphi(\cdot)$ with L_2 regularized MLP layers,

$$Z_l = g_\varphi(H_l), Z_a = g_\varphi(H_a), \tag{2.20}$$

where φ represents the parameters of the projector $g_\varphi(\cdot)$.

Finally, the contrastive loss $\mathcal{L}$ is employed to maximize the consistency between the projections $z_{l,i}$ and $z_{a,i}$. To achieve this, a symmetric temperature-scaled cross-entropy loss is utilized to improve representation consistency between the two views. This loss function, by adjusting the temperature parameter, regulates the influence of negative samples, enabling the model to more effectively learn the strong relationships between similar nodes:

$$\mathscr{L} = \frac{1}{2n} \sum_{i=1}^{n} \left[\ell\left(z_{l,i}, z_{a,i}\right) + \ell\left(z_{a,i}, z_{l,i}\right) \right],$$

$$\ell\left(z_{l,i}, z_{a,i}\right) = \log \frac{e^{\frac{\sin\left(\mathbf{z}_{l,i}, \mathbf{z}_{a,i}\right)}{t}}}{\sum_{k=1}^{n} e^{\frac{\sin\left(\mathbf{z}_{l,i}, \mathbf{z}_{a,k}\right)}{t}}} \tag{2.21}$$

During the fixed number of learning iterations, we update the anchor view A_a using the gradually evolving enhancements of the learned S:

$$\mathscr{A}_a \leftarrow \tau \mathscr{A}_a + (1 - \tau)\mathscr{S}, \tag{2.22}$$

where $\tau \in [0, 1]$ is the decay rate of the anchor graph. Typically, τ is set greater than 0.99 to ensure the stability of the training process.

2.5 Performance of Denoising Framework

In this section, we employ the BMSD framework to perform structural denoising on tabular behavior molecule data, aiming to further optimize its structural information. The experiments in this chapter assess the proposed method's effectiveness by training multiple machine learning models on tabular datasets and analyzing their performance.

2.5.1 Analysis of Classification of Experimental Results

Specifically, we use the behavior molecule dataset obtained through behavioral modeling as the input to the BMSD framework to obtain optimized, high-quality behavior molecule graph data. In the classification task, we select the graph isomorphism network (GIN) as the primary model for graph learning, which is employed for representation learning and classification performance validation on the optimized behavior molecule graph data. This process evaluates the BMSD framework's effectiveness in optimizing behavior molecule data and performing classification tasks while examining its generalization across various models and datasets.

Table 2.1 summarizes the average performance scores on real-world datasets. The experimental results clearly show that the BMSD framework outperforms all other methods across datasets, validating its effectiveness and robustness in

Table 2.1 Comparison of classification task results. The highest results are highlighted in *bold*

Metric	Acc	Rec	F1-score
Dataset	*Credit approval*		
KNN	0.680±0.024	0.665±0.030	0.665±0.027
GaussianNB	0.796±0.022	0.765±0.014	0.774±0.019
BernoulliNB	0.793±0.036	0.788±0.035	0.786±0.035
DecisionTreeGini	0.850±0.017	0.858±0.013	0.847±0.016
DecisionTreeEntropy	0.818±0.016	0.811±0.013	0.811±0.015
SVM	0.717±0.024	0.679±0.016	0.682±0.020
RandomForest	0.873±0.023	0.866±0.019	0.868±0.022
XGBoost	0.855±0.015	0.848±0.011	0.849±0.014
Catboost	0.851±0.021	0.858±0.017	0.849±0.021
LightGBM	0.865±0.023	0.860±0.022	0.860±0.022
DANet	0.844±0.025	0.846±0.025	0.843±0.026
BMSD-N	0.921±0.013	0.930±0.017	0.919±0.016
BMSD	**0.942±0.009**	**0.940±0.015**	**0.941±0.010**
Dataset	*Heart Disease*		
KNN	0.486±0.038	0.203±0.031	0.174±0.035
GaussianNB	0.522±0.058	0.315±0.058	0.290±0.059
BernoulliNB	0.542±0.038	0.270±0.033	0.249±0.037
DecisionTreeGini	0.538±0.046	0.200±0.000	0.140±0.008
DecisionTreeEntropy	0.486±0.044	0.294±0.037	0.289±0.029
SVM	0.575±0.056	0.303±0.052	0.285±0.056
RandomForest	0.573±0.053	0.305±0.048	0.287±0.041
XGBoost	0.539±0.038	0.290±0.060	0.275±0.054
Catboost	0.572±0.031	0.273±0.041	0.241±0.042
LightGBM	0.241±0.042	0.262±0.036	0.242±0.042
DANet	0.486±0.038	0.203±0.031	0.174±0.035
BMSD-N	0.783±0.017	0.665±0.015	0.644±0.018
BMSD	**0.810±0.011**	**0.702±0.014**	**0.680±0.015**

optimizing behavior molecule graph structures. At the same time, we observe that since both the contrastive view and anchor view in the BMSD framework are derived from the initial attribute score matrix, the structural denoising improvement achieved through contrastive learning is limited.

This observation provides an important insight for our future research direction: optimizing the construction of the contrastive view may be key to further enhancing the performance of the BMSD framework. Specifically, this issue falls within the scope of behavior molecule modeling methods, as generating more suitable contrastive views from the original tabular data is one of the core challenges of behavior molecule modeling. By designing high-quality contrastive views and incorporating graph contrastive learning, we can more effectively preserve critical information from behavior molecule graphs generated by different types of behavior molecule modeling methods, thereby removing noise to a greater extent, optimizing structure, and generating higher-quality behavior molecule graph data.

2.5.2 *Visualized Analysis*

To visually demonstrate the behavior molecule data optimized through the BMSD framework, we provide four examples in Fig. 2.18. These examples compare the results obtained using only behavior molecule modeling methods with those obtained after denoising optimization. Significant structural changes can be observed from the comparisons, offering critical evidence for analyzing the effectiveness of the BMSD framework.

Initially, behavior molecule graphs constructed directly using behavior molecule modeling methods often exhibit sparse structures. This sparsity primarily arises from the reliance on attribute correlations during the graph construction process, where correlation scores between attributes are typically low. When a fixed connectivity threshold is applied, many potential connections are excluded, resulting in a sparse adjacency matrix.

In contrast, behavior molecule graphs optimized through the BMSD framework achieve a balanced trade-off between structural density and representational capability. This improvement is attributed to the BMSD framework's contrastive learning

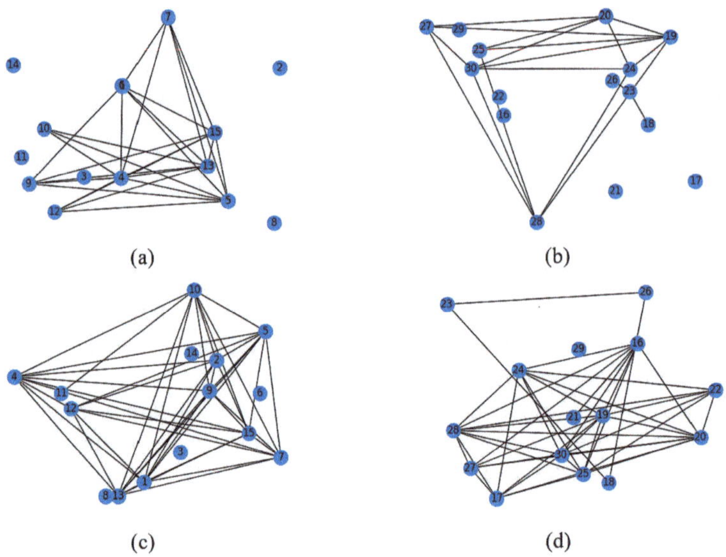

Fig. 2.18 Visualizations of the credit approval dataset under different models: (**a**) and (**b**) represent behavior molecule graphs modeled without denoising, while (**c**) and (**d**) depict behavior molecule graphs optimized and denoised using the BMSD framework

mechanism. Specifically, the anchor view provides a global structural reference derived from the original data, while the contrastive view undergoes continuous adjustments during optimization. By optimizing the consistency between these two views, the framework extracts shared essential structural information while discarding noisy and inconsistent elements.

The resulting behavior molecule graph retains critical relational patterns while effectively removing redundancy and noise, yielding a graph structure characterized by low noise and high representational capacity. This comparison illustrates the pivotal role of the BMSD framework in optimizing behavior molecule data. It not only enhances the structural quality of the data but also provides superior input for downstream tasks, thereby significantly improving the performance and robustness of machine learning models.

2.6 Conclusion

In this work, we introduce a novel perspective to elucidate the behavioral structure by capturing the interactions between behavioral attributes. Specifically, we propose a tangible graph structure for behavior description, which can reflect the uniqueness and stability of behaviors. To achieve this, we design the so-called behavioral molecular structure model that generates attribute representations rich in behavioral information. In this model, we analogize the attribute interactions to atomic bonds, forming a comprehensive set of relationships to describe behavioral structure. We then employ graph representation learning algorithms to extract rich behavioral information, which serves as the foundation for constructing highly expressive behavior models. We validate the effectiveness of our proposed method on the different representative downstream tasks: behavior detection, behavior prediction, and behavior generation. Experimental results demonstrate that utilization of the interactions of atomic-level attributes in behavior significantly improves the performance of these downstream tasks. Furthermore, we provide an initial explanation for its performance gains from an expressive power perspective. Overall, we employ behavior molecular structure to provide an expression of behavioral structure, aiming to propose promising signals for the deep utilization of behavioral information. It calls for the research community to increase awareness and lay the groundwork for future research on behavioral structure.

Meanwhile, we propose a novel method to optimize the behavior molecule graphs, termed the behavior molecular structure denoising method. The method seeks to optimize graph structures by aligning the learned structure with a thoughtfully designed self-supervised learning objective. Leveraging an unsupervised graph contrastive learning strategy, BMSD effectively uncovers and exploits latent information within the data. It not only guides the model to discover hidden, informative connections but also significantly reduces noise interference in the original graph structure. This process improves the model's capacity to capture underlying structural patterns, providing higher-quality input data for downstream tasks. Experimental results demonstrate that behavior molecule graph data optimized and denoised by BMSD consistently achieve superior performance across various real-world applications. The optimized graph structures more accurately capture critical patterns in the original data while mitigating the negative impacts of noise on model training.

References

1. T.N. Kipf, M. Welling, arXiv preprint arXiv:1609.02907 (2016)
2. C. Shang, Q. Liu, Q. Tong, J. Sun, M. Song, J. Bi, Neurocomputing **445**, 12 (2021)
3. D. Bo, X. Wang, C. Shi, M. Zhu, E. Lu, P. Cui, in *Proceedings of the Web Conference 2020* (2020), pp. 1400–1410
4. A. Hasanzadeh, E. Hajiramezanali, K. Narayanan, N. Duffield, M. Zhou, X. Qian, in *Advances in Neural Information Processing Systems 32: Annual Conference on Neural Information Processing Systems 2019, NeurIPS 2019*, vol. 32, pp.10711–10722 (2019)
5. N. Mehrabi, F. Morstatter, N. Saxena, K. Lerman, A. Galstyan, ACM Comput. Surv. (CSUR) **54**(6), 1 (2021)
6. C. Song, Z. Qu, N. Blumm, A.L. Barabási, Science **327**(5968), 1018 (2010). DOI 10.1126/science.1177170. https://www.science.org/doi/abs/10.1126/science.1177170
7. N. Goyal, H.V. Jain, S. Ranu, in *Proceedings of The Web Conference 2020* (2020), pp. 1253–1263
8. B. Hanin, D. Rolnick, in *International Conference on Machine Learning* (PMLR, New York, 2019), pp. 2596–2604
9. L. Franceschi, M. Niepert, M. Pontil, X. He, in *International Conference on Machine Learning* (PMLR, New York, 2019), pp. 1972–1982
10. Y. Chen, L. Wu, M. Zaki, Adv. Neural Inf. Proces. Syst. **33**, 19314 (2020)

Chapter 3
Behavior Molecule-Level Learning by Behavioral Interior Reconstruction

Abstract Detecting deceptive behavior has become a critical challenge as malicious behaviors grow more sophisticated. Cunning attackers (e.g., telecommunication fraudsters and insider spy employees) confuse their deceptive behaviors with benign behaviors, which has rendered traditional detection methods increasingly indistinguishable. By adopting a mindset that maps behavior onto a high-dimensional attribute space, valuable clues can be derived to aid in detecting deceptive behavior by transforming nearly inseparable behavioral data into separable patterns within a behavioral attribute space. Within the current landscape, the work of effectively learning behavioral attributes remains largely unknown. This work proposes a novel method called *mo*tif-based behavior *re*construction (*MORE*). It reconstructs behavior as the frequent connectivity pattern on the granularity of attributes instead of representing behaviors, where we design a transformation network for learning a vector field to get expressive attribute representations. We assess our work across four representative security scenarios: detecting fraudulent transactions in online payment services, identifying malicious domains in domain name systems, detecting insider threats within organizational information systems, and identifying suspicious traffic within anonymous network communications. These experiments are conducted using private and public datasets, as well as real and simulated data that encompass various deceptive behaviors relevant to different security domains.

3.1 Introduction to Behavioral Interior Reconstruction

Security incidents are on the rise in modern information systems, and defenders face attacks with increasing sophistication [1, 2]. Deceptive behaviors, in security incidents, may result in critical security impacts or even system breakdowns. Answering with high confidence whether a behavior is deceptive is however challenging since that security incidents often stem from ambiguity, insufficient specification, and logical defects in complex business rules [3, 4]. Meanwhile, attackers are constantly deploying a myriad of tactics to confuse defenders and bypass the security solutions they have implemented. As an example, attackers

C. Wang, *Computational Structural Behavior*,
https://doi.org/10.1007/978-981-95-1013-9_3

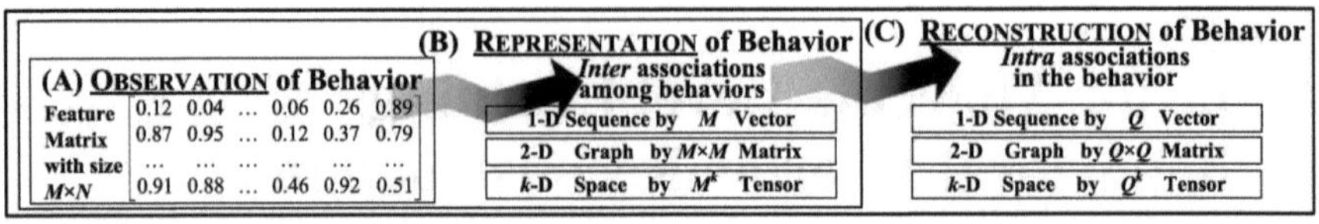

Fig. 3.1 The illustration of understanding behavior mathematically involves transitioning from observation to representation and to reconstruction

often employ tactics such as prolonged lurking within the target system to conceal their identity, intentionally avoiding the reuse of attack materials, and tend to favor specific virtual hosting with loose policies (e.g., not requiring real online identities). Thus, there is a need to revisit the technologies used for identifying deceptive behaviors to ensure accurate and timely prevention of security incidents.

Existing Solutions It is a triumph to reduce great volumes of recording operations to the low-dimensional description consisting of just a few defined attributes [5]. The observation of behavior (as shown in Fig. 3.1a) provides a window into understanding the intentions behind behaviors [4, 6]. This fact makes it more or less possible to use machine learning methods to analyze behaviors and their underlying intentions. For instance, many security tasks (e.g., fraud detection and malicious domain detection) can be formulated as a classification problem [7–9].

Attackers adopt a series of strategies to mimic benign users and generate "seemingly benign" behaviors (e.g., frequently changing attack methods to avoid blacklists, simulating benign behavioral activities to confuse authentication schemes, and even launching gang-level attacks via the black market exchange) [10, 11]. The deceptive behaviors are unknowingly concealed by attackers among similar benign behaviors, which leads to confusion in implementing security solutions. All the strategies, which directly affect the observation of behavior, aim to show benign intentions to defenders while, in reality, committing attacks on the target.

Advanced neural network methods can seek a clear gap in the hidden space to distinguish different behaviors, where each behavior is typically transformed into a representation [12–14]. They distort and deform the original input of behaviors through affine and nonlinear transformations until different behaviors can be easily distinguished from others. Recent researches show a promise in detecting entangled anomalies (e.g., fraud and network attack) through the analysis of associations among behaviors [15, 16]. Researchers exploit the sequence, graph, and even higher-dimensional space (as shown in Fig. 3.1b) to model the associations among behaviors [17]; it forces attackers to carefully erase invisible camouflage traces, e.g., traffic access from similar IP addresses and APPs from the same malware family.

In the ongoing cat-and-mouse game between attackers and defenders, the clues available to defenders are becoming increasingly elusive, while the cost of differentiating between deceptive and benign behaviors continues to rise. Even if attackers are successfully blocked, defenders get fatigued with huge costs in computability. In extreme cases, attackers make subtle deviations from benign behaviors, so that deceptive behaviors surround benign ones (like links in *knot theory* [18]). It is difficult for defenders to separate such a group of confusing behaviors. Stuck in the confines of a coarse-grained approach to directly analyzing behaviors, a straightforward response is to pull apart the behaviors at the cost of minimal overlap. Obviously, this is a rudimentary solution but not the ideal solution to be pursued.

Behavioral Structure Fortunately, the classical literature on topology [19, 20] reminds us that behavior with even a small number of attributes could encode a high-dimensional structure representing the connectivity/associations of attributes (as shown in Fig. 3.1c). We recognize that current research efforts in identifying deceptive behaviors primarily focus on sharing and absorbing information related to similar behaviors (i.e., behaviors with similar contexts). By solely studying behavior correlations, attackers may become aware of these solutions, leading to subpar performance.

In this work, we represent the interconnections among all behavioral attributes as connectivity patterns in networks where the behavior can be known as the *motif*[1] (i.e., local structures composed of some behavioral attributes and their interconnections). Along this way, we introduce structural information to separate a set of entangled behaviors. Compared with the representation of behavior, it is closer to the minimum information dimension required to predict all behaviors completely with the aid of observed behaviors. We reconstruct behaviors on the granularity of attributes and employ neural networks to learn a vector field that can transform the attributes according to the context of attributes in behaviors. Instead of seeking the best segmentation gap to distinguish different behavior representations, we learn more expressive attribute representations for each behavior, which can leave deceptive behaviors nowhere to hide. Defenders can perceive all observable behaviors, where the advantage in behavior perception ensures that attackers' tricks are difficult to confuse the semantics of behavioral attributes.

Proposed Solution We propose *MORE*, the first work to identify deceptive behavior by capturing the behavioral inconsistency at motif level. MORE comprises three components: the topology of behavior, the representation on behavior topology, and the detection of deceptive behavior. MORE constructs a customized graph on the attribute space to model the associations among behavioral attributes. Then, we correspondingly develop a semi-supervised representation learning mechanism to transform the attributes into expressive representations in the context of observed behaviors. Meanwhile, it learns from the observed behaviors in a global way, which maintains the improvement brought by the representation of behavior, namely, the associations among behaviors. Finally, with expressive attribute representations, MORE characterizes the behavior via customized features for detecting deceptive behaviors.

[1] Motifs are defined as the patterns of interconnections that recur in many different parts of a network at frequencies much higher than those found in randomized networks [21, 22].

Our extensive evaluation shows that our work is capable of capturing deceptive behaviors with high performance and low misreport rate. We also reveal some interesting findings from the evaluation, which can be viewed as an initial step toward understanding behavioral structure.

Contributions Our contributions are threefold:

- ***New awareness.*** We revisit existing security techniques and propose a new solution—*behavior reconstruction.* To raise security community awareness of behavioral structure, we advocate for the utilization of high-dimensional behavioral structures at the attribute level, which can provide defenders with an information advantage.
- ***Novel solution.*** We design *MORE* for encoding behaviors as motifs (i.e., connectivity patterns on behavioral attributes) and transform behavioral attributes into expressive representations in an effective way for detecting deceptive behavior.
- ***Valid effects.*** We perform an extensive evaluation on the effectiveness of MORE in *multiple security scenarios*. The results show that MORE outperforms existing security solutions in detecting deceptive behaviors. We also demonstrate the usefulness of MORE to analyze deceptive behaviors and categorize behaviors into different clusters in low-dimensional visualization.

3.2 Beyond Feature-Based Detection

Over the past few years, researchers proposed several methodologies to detect deceptive behaviors based on observed features [15, 23], e.g., statistical features computed from the transaction responses [24], lexical features containing typosquatting and obfuscation [25], and third-party features derived from public ratings [26]. Many works analyze the suspicious domains based on the lexical features or URL ranking to determine the deception of the site [27]. As an intuitive solution, many existing works draw support from classification algorithms to inspect deceptive behaviors [8, 28, 29]. The existing typical methods (such as linear-based methods and tree-based methods) are good at distinguishing the representations to get two separable regions when facing this situation. With the fast evolution of attackers' means, the rigidity of feature-based solutions extends the gap in successfully mitigating deceptive attacks.

It is challenging to separate deceptive behaviors from behavioral data with the large-scale implementation of crafted tricks in attacks [10, 30]. Many researchers realize that deep learning methods can transform behaviors as the representations in a theoretically distinguishable space through complex network design [31–33]. Li et al. [34] modeled the normal patterns inside data through hierarchical variational autoencoder with two stochastic latent variables, each of which learns low-dimensional inter-metric or temporal embeddings. Meanwhile, there are an increasing number of studies that focus on modeling behavior as the edge to express interactions between users [35, 36], e.g., using the bipartite graph to describe trading networks [37]. Zhang et al. [38] introduced an attributed heterogeneous information network for user representation to build up relatedness over users in underground forums. Li et al. [36] introduces a dual-domain graph convolutional network for insider threat detection, considering both user behaviors and relationships. The representation of behavior and their associations (e.g., sequence and graph) has proved useful in detecting deceptive behaviors in security scenarios [4, 39]. Zeng et al. [1] abstracted high-level behaviors from audit logs by aggregating semantic information from different events. Liu et al. [40] took into account causal and sequential information by adopting a rule-based approach to map relationships among log entries, where the behaviors are reflected in a graph. The intuition behind this is that attackers are hard to fake nonintuitive interactions with other behaviors. Most graph-based detection algorithms focus on finding dense blocks, discovering local subgraphs, designing belief propagation, and latent factor models [12, 41]. Essentially, they utilize the homophilic effect exhibited by graphs to detect deceptive behaviors as node classification. Unfortunately, as attackers employ cunning tactics (such as making their behaviors capricious, fragmented, and imperceptible), it is easy for them to hide the perceptible associations among behaviors.

3.3 Background

3.3.1 Scenario

Since the implementations of information systems are various in security applications, we focus on selected typical behaviors that consist of the processing attributes/features to make maximally informative descriptions of behavior.

Fraudulent Transaction Detection in Online Payment Services Online payment service providers face serious threats from deceptive behaviors (e.g., telecommunications fraud, phishing websites, and coupon clipping in gangs). There is an urgent need to realize effective and comprehensive fraud detection [39]. The deep methods are popular since they mine the relationships between behaviors. They have made considerable progress in detecting deceptive behaviors. Unfortunately, attackers are starting to change their tactics, becoming invisible to evade detection systems. Some of them try to look benign by adopting a strategy of throwing a stone to clear the road (i.e., they usually perform several benign actions before revealing their attacking intents) [42]. Some of them incite users to participate in negligently set promotional rush buying activities to cover up their behaviors [38]. Even worse, these tricks are cheap and easy to use and allow attackers to successfully circumvent regulations.

Malicious Domain Detection in Domain Name Systems Attackers can abuse the domain name system (DNS) to lure users to be victims of malicious domains hosting drive-by-download malware, botnets, phishing websites, or spam messages [27].

For instance, attackers mimic the looks of legitimate websites to create phishing domains and leverage social engineering techniques to trick users into clicking deceptive links. Attackers also randomly send mails to widely used mailbox names within a domain, such as "info," "admin," and "support," to seek the valid email addresses.

Insider Threat Detection in Organizational Information Systems Attackers initially compromise the accounts and hosts of insider employees who can intentionally use authorized access to influence the confidentiality, integrity, or availability in the organizational information system [1, 40]. Then, attackers can stealthily and persistently compromise multiple hosts through the intranet and steal confidential information. In such scenarios, we can obtain a comprehensive behavioral dataset that contains relatively complete records of users' multiple behaviors (e.g., log-in operations, removable storage device usage, file operations, network operations, and email traffic).

Suspicious Traffic Detection in Anonymous Network Communications The darknet can offer notable hidden services in the underground marketplace. So many worldwide users are connected via bridge to hidden services (such as Tor), where about 20% of these services are suspicious [25, 43]. Analyzing darknet traffic helps in the detection of deceptive activities. For example, illegal trading often tops the hidden services list with the highest number of closely related services.

3.3.2 Motivating Example

Here, we present two examples to illustrate the limitations of existing work. The examples come from actual data from online payment services and synthetic data from organizational information systems.

Example 1 False online shopping transactions. Scammers gain access to the victim's account or deceive them into making fraudulent transfers. Scammers avoid providing their own personal accounts to receive the stolen funds but instead mimic benign behaviors in daily transactions, purchasing similar items from fake merchants acting as money launderers. This strategy aims to camouflage fraudulent transaction behaviors as part of regular shopping activities, misleading analysts into labeling it as benign behavior.

Example 2 Unusual overtime activities before resignation. An employee who previously did not work after regular hours starts accessing the organization's internal systems after hours. He/she copies data using a removable drive or uploads data to third-party websites and, shortly thereafter, leaves the company. Each action of this employee appears harmless but actually exposes sensitive business data. It is challenging to detect such deceptive behaviors before related security incidents occur.

Current machine learning methods that rely on observation and representation can identify tiny differences in fixed dimensions. However, they are increasingly struggling when facing well-planned attacks from deceptive actors that involve carefully orchestrated behaviors in multiple dimensions (especially when seemingly benign behaviors combine to present deceptive intent).

The superiority of our method lies in the fact that the differences in the context of behaviors (such as semantic associations among behavioral attributes) are hard for attackers to sense and disguise. Defenders with the information advantage can see and exploit those associations, which will lead to a qualitative improvement in detecting deceptive behaviors. For instance, in transaction scenarios, these fraudulent accounts often have underlying correlations as they are typically created by fraudsters engaging in a large volume of business. Defenders can observe all fraudulent transaction data, which collectively point to behavioral attributes highly correlated with detected fraudulent transactions, and this correlation is imperceptible to the fraudsters. In the context of insider threat detection, employees attempting to avoid detection first engage in operations during off-peak hours when staff is scarce and then discreetly transfer sensitive data they have access to using local or cloud storage. In normal business operations, certain business data should theoretically not be accessed during certain time periods (such as after hours). The inappropriate timing combined with frequent access operations is a clear anomaly signal.

3.4 Motif-Based Behavior Reconstruction

We first give the high-level idea of MORE in Fig. 3.2 and provide a problem formulation for detecting deceptive behaviors, followed by detailing its technical compositions. MORE consists of the following parts: (A) MORE records the behavior of benign users and malicious attackers within the information system through monitors and stores behavioral data as behavioral attributes. (B) Next, Sharper constructs a behavior subgraph for each behavior and generates a behavior graph based on the corresponding behavior subgraphs for all behaviors. (C) The

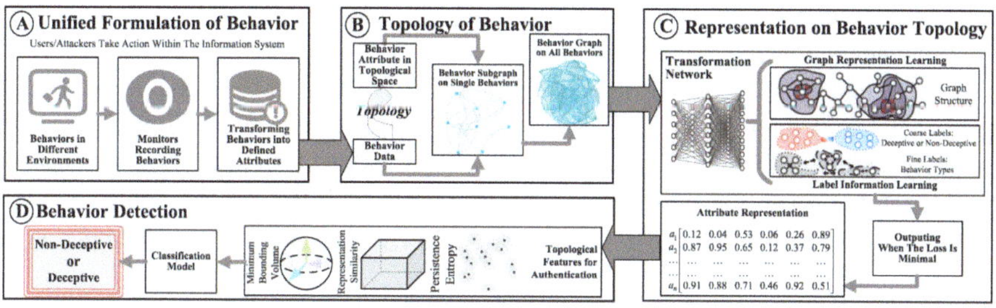

Fig. 3.2 The overview of MORE

behavior graph is fed into the transformation network for learning the graph structure and label information, and the attribute representation corresponding to each behavioral attribute is obtained. (D) MORE generates customized topological features based on attribute representations and applies classification models to determine whether the behavior is deceptive.

3.4.1 Unified Formulation of Behavior

In security applications, diverse monitoring solutions are deployed to reduce the impact of deceptive behaviors and ensure the normal operations of information systems. The behavioral data sources are scattered in different environments, which leads to the nonuniformity of collected behavioral data in terms of acquisition standards (e.g., different fields, formats, and timing of collection). To store and analyze data more effectively, the initial behavioral data are required to use basic expertise to transform the captured behaviors into well-defined discrete attributes. For continuous attributes such as time and amount, we discretize them into bins. It is important to note that the binning intervals have an impact on the sensitivity of the method. However, high sensitivity does not necessarily enhance the detection capabilities of our method. In fact, it may lead to the loss of necessary associations, such as partitioning close values into different intervals. Let $\mathscr{B}$ be a set of behavioral data. Then, $B \in \mathscr{B}$ represents a behavior in a form of $B = \{f_1, f_2, \cdots, f_{|B|}\}$, where f_i is the ith attribute of B and $|B|$ is the number of attributes in B. Optionally, we can use information gain to select the fields in the raw data that have the highest impact on detecting deceptive behaviors as behavioral attributes. In the case of missing values for some behavioral attributes, the corresponding missing elements are removed from set B.

Based on the prior feedback and domain rules, a part of behaviors can be coarsely marked as deceptive or trustworthy and finely marked as specific behavior types. Let $\mathscr{Y}^c$ and $\mathscr{Y}^f$ as the set of coarse-grained labels and fine-grained labels for $\mathscr{B}$, respectively and y_i^c and y_i^f are the labels of the ith behavior B_i, where $y_i^c \in \{-1, 0, 1\}$ and y_i^f belongs to the set of nonnegative integers. For the unlabeled behavioral data, we mark their labels as 0, i.e., $y_i^c = 0$ and $y_i^f = 0$. In $\mathscr{Y}^c$, $+1$ represents deceptive behavior, and -1 represents nondeceptive behavior. In $\mathscr{Y}^f$, each integer represents a behavior type.

Given the behavior B, our objective is to identify whether the behavior is deceptive or not and then classify it into the appropriate behavior type for subsequent security analysis. In light of the above, we define the detection of deceptive behavior as the estimation of the probability $(\mathscr{Y}^c, \mathscr{Y}^f|\mathscr{B})$, which represents the likelihood of B being inconsistent with regulations and correctly identified as behavior type y^f.

3.4.2 *Topology of Behavior*

To analyze the inconsistency of deceptive and trustworthy behaviors effectively, the structure of behavior is required to characterize behavioral data in a topological manner. Instead of using the feature matrix or behavior representations, we propose a new solution based on behavioral attributes to characterize behaviors from the topological perspective.

3.4.2.1 Topological Space

Given that a behavior B is a nonempty set, we can denote $\mathscr{T}$, a subset of the power set of B (or a set family composed of some subsets of B), as a topology of B in the light of the topology definition. Accordingly, we get the topological space $(B, \mathscr{T})$ based on the behavior B and its topology $\mathscr{T}$. Generally, a set can specify many different topologies. It is necessary to specify both the set and the specified topology, when talking about a topological space. So the problem is how to select the subsets of B to generate the topology $\mathscr{T}$.

We first add the empty set $\emptyset$ and the set B into $\mathscr{T}$. Considering the definition (the intersection of a limited number of elements is still in $\mathscr{T}$), we decompose the behavior B into a series of independent elements denoted by f_i, where i represents the integer within the range from 1 to the cardinality of B. For each element f_i, we add it into $\mathscr{T}$. Then, we select the subset of B, consisting of two or multiple elements, and add it to $\mathscr{T}$ in a bottom-up way. In this way, the axiom (the union of any number of elements in $\mathscr{T}$ is still in $\mathscr{T}$) is satisfied. Specifically, we first select two elements to form a subset of cardinality two. Then, based on independent elements and subsets of cardinality two, we can generate subsets of cardinality three or four by combining them. Finally, we can iteratively generate a subset of cardinality $|B|$ and get the final $\mathscr{T}$.

We simplify generating the behavior topology $\mathscr{T}$ to how to select the appropriate subsets of cardinality two. Once these subsets are determined, we get the corresponding topology $\mathscr{T}$ and topological space $(B, \mathscr{T})$ by our above operations. Next, we have customized a behavior subgraph to help us choose which two elements should form a subset of B.

3.4.2.2 Behavior Subgraph on Single Behavior

Given a behavior B and a topological space $\mathscr{T}$, we can abstract *simplicial complexes* as the high-dimensional generalization of the behavior subgraph. Once the nodes (i.e., the behavioral attributes in B) are known, the geometric simplicial complex can be fully characterized by the combinatorial description of a collection of *simplices* satisfying the incidence rules. In topology, the graph is the simplicial complexes of dimension one. The combinatorial description of any simplicial complex gives rise to an abstract simplicial complex that can be seen as topological spaces, and the corresponding geometric complexes can be seen as geometric realizations of their underlying combinatorial structure. So we treat the graph as a topological space from which the topological properties can be inferred.

A simple and feasible method is an immediate extension of the notion of the d-neighboring graph.[2] Here, the parameter d can be viewed as a resolution of the data. Our behavior subgraph changes as d increases where new connected components might appear, the existing components might merge, the loops and cavities might appear or be filled, and so on. We track these changes, identify the discriminative features, and reveal the structure of behavior. In a recorded behavior B, the co-occurrence of attributes is a phenomenon that cannot be neglected. If parameter d is too small (e.g., an extreme setting of d is 0), there is no direct or indirect co-occurrence relationship between any two attributes, which is meaningless to restore the behavior itself. On the contrary, if d is too large (e.g., positive infinity), it will make all the attributes coexist directly, and it is impossible to distinguish deviation. Therefore, we need to set the value of d that satisfies any two attributes have a direct or an indirect co-occurrence relationship. In other words, in the behavior subgraph, any node must be the neighbor node of other nodes (first-order or higher-order neighbors).

In this work, we introduce the euclidean distance to compute the distance, denoted as $distance(f_i, f_j)$, between any two attributes f_i and f_j in the metric space. In fact, several methods can also calculate the similarity between two vectors; we only demonstrate the use of euclidean distance here. The distance $distance(f_i, f_j)$ between two attributes f_i and f_j is defined as the length of the segment connecting the two points in euclidean space, where the recorded position corresponding to the point can be recorded as a vector. Obviously, the distance between any two attributes is directly influenced by their vectors. Therefore, we need to set an appropriate euclidean space to reflect the practical significance, which is time-consuming and depends on external expert knowledge in special scenarios. Alternatively, we adopt the widely used pre-trained vectors to determine the location of attributes in the space. The pre-trained word vectors, e.g., fastText, can be obtained in many common corpus data with strong generality and then be fine-tuned or directly applied. For any two attributes f_i and f_j, we denote their corresponding pre-trained word vectors (or the positions in euclidean space) as $\vec{\mathbf{f_i}}$ and $\vec{\mathbf{f_j}}$, respectively. Then, we can calculate the distance $distance(f_i, f_j)$ by

$$distance(f_i, f_j) = ||\vec{\mathbf{f_i}} - \vec{\mathbf{f_j}}||. \tag{3.1}$$

Furthermore, we provide the procedure of generating behavior subgraph in Algorithm 1. First (Steps 1–2), we initialize the distance matrix M to be positive

[2] Given a set of points $\{x_0, \cdots, x_k\}$ in a metric space and a parameter $d \geq 0$, the Vietoris-Rips complex $Rips_d$ is the set of simplices $\{x_0, \cdots, x_k\}$ such that the distance $distance(x_i, x_j)$ between any pair of points x_i and x_j is less than or equal to d for all (i, j).

infinity and set the adjacency matrix of all 0 and all 1 to facilitate subsequent calculation. Obviously, the distance between two same nodes is 0, so we set the diagonal elements of M to 0 in Steps 3–4. We calculate all the distances between any two behavioral attributes in a behavior and put each distance into the empty set to get $Dlist$ in Steps 5–7. In particular, we sort the $Dlist$ in ascending order, which ensures that subsequent invalid computations can be avoided. Only when the distance value is increased to the value in $Dlist$ does the structure of the behavior subgraph change. So the value of parameter d in the d-neighboring graph must be an element in $Dlist$. For each element d' in $Dlist$, we calculate the connectivity of its corresponding graph structure to determine whether any two attributes have the co-occurrence relationship (Steps 9–16). Note that parameter d is used to help us generate behavior subgraphs from each behavior. Therefore, in our algorithm, the value of d is not returned; only the adjacency matrix representing the behavior subgraph is returned.

As shown in Fig. 3.3, we present five snapshots of generating behavior subgraph corresponding to a behavior consisting of eight attributes. In Fig. 3.3a, only the two most similar node pairs are selected. As d increases, more and more nodes are connected. Until updating to the T16 snapshot, we obtain a connected graph. In this case, we can see that two attributes connected in the last time are semantically similar but significantly different from the other six attributes.

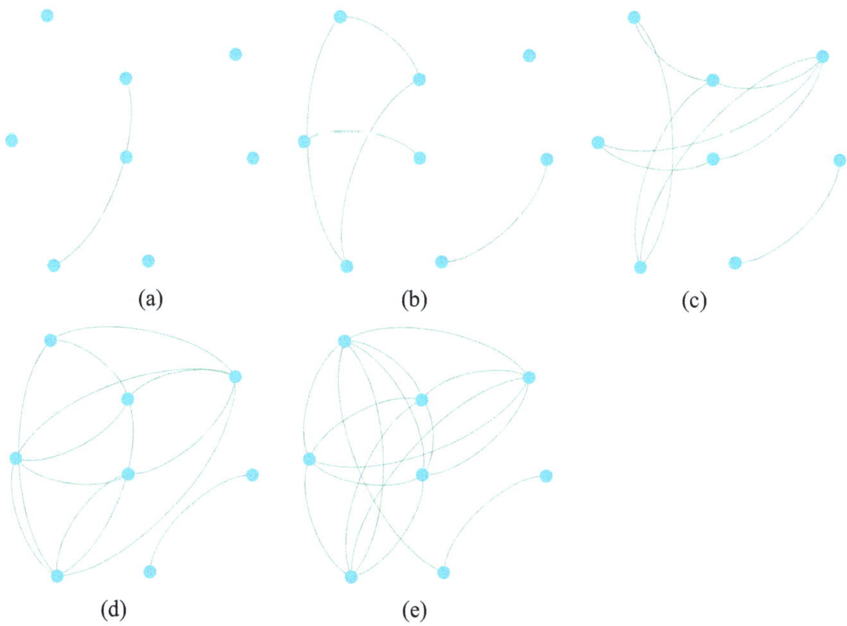

Fig. 3.3 A toy example of constructing the behavior subgraph. (**a**) T0, (**b**) T4, (**c**) T8, (**d**) T12, (**e**) T16

Algorithm 1: Generating behavior subgraph

Input: The behavior B
Output: The adjacency matrix of behavior graph M_b

Initialize distance matrix $M = [+\infty]_{|B|\times|B|}$, where $|B|$ is the cardinality of B;
Initialize adjacency matrix $M_1 = [1]_{|B|\times|B|}$ and $M_0 = [0]_{|B|\times|B|}$;
foreach $i \in range(0, |B|)$ **do**
  $M[i, i] = 0$;
end
foreach $i \in range(0, |B|)$ **do**
  foreach $i \in range(0, |B|)$ **do**
    $M[i, j] = GetDistance(B[i], B[j])$;
  end
end
Get $Dlist$ by sorting all elements in M in ascending order;
foreach $d' \in Dlist$ **do**
  $M_b = numpy.where(M <= d', M_1, M_0)$;
  $M_b' = numpy.where(M <= d', M_1, M_0)$;
  foreach $i \in range(2, |B|)$ **do**
    $M_b' = M_b \times M_b'$;
    $M_b' = numpy.where(M_b' >= 1, M_1, M_0)$;
    if $\sum(M_b') == |B|^2$ **then**
      Return M_b
    end
  end
end

3.4.2.3 Behavior Graph on All Behaviors

According to the description of knowledge graph, we define the behavior graph as a set of numerous co-occurrence triples in the form of Resource Description Framework graph [16, 44]. More specifically, the behavior graph is a graph composed of multiple behavior subgraphs that contain shared nodes. Each behavior subgraph corresponds to a behavior B and is composed of several attribute-level co-occurrence triplets. Each triplet, corresponding to the associations between two behavioral attributes, consists of three elements $(f_i, r_{i,j}, f_j)$ that codify the co-occurrence relationship $r_{i,j}$ of two attributes (f_i and f_j) in the form of $(Header, Relation, Tail)$. The relationship means that two attributes (f_i and f_j) appear together in the behavior B. $Relation$ depends on the type of f_i and f_j and means the co-occurrence type between two specified attributes. Note that $Head$ and $Tail$ cannot have the same attribute type, e.g., two different source IP addresses cannot occur in a behavior simultaneously.

We formalize our behavior graph as $G = (V, E, R)$, where V is the node set that contains all potential values in behavioral attributes, E is the edge set that contains co-occurrence triplets generated from all behavioral data, and R is the relation set that contains all potential relation types. Since a series of triples can be constructed based on a single behavior in Sect. 3.4.2.2, we define a behavior subgraph G' on the above triples. In practice, the relationship between any two attributes is not only directly dependent on the behaviors that contain them but also indirectly dependent on other behaviors that do not contain them. In this case, behavior graph G is composed of all behavior subgraphs, where the node set of G is the union of all subgraph node sets, the edge set is the union of all subgraph edge sets, and the relation set is the union of all subgraph relation sets. Based on behavior set $\mathscr{B}$ of size $|\mathscr{B}|$, we denote the behavior graph as

$$G = (V, E, R) = \left(\bigcup_{i=0}^{i=|\mathscr{B}|} V_i', \bigcup_{i=0}^{i=|\mathscr{B}|} E_i', \bigcup_{i=0}^{i=|\mathscr{B}|} R_i' \right),$$

where V_i', E_i', and R_i' are the node, edge, and relation set of the behavior subgraph G_i' for each behavior $B_i \in \mathscr{B}$.

3.4.3 *Representation on Behavior Topology*

We present the details of the transformation network in MORE. We transform the input behavior set $\mathscr{B}$ and behavior graph G into expressive attribute representations, which allows us to find the deviation of behavior. The transformation network is designed in order to achieve two goals:

- Graph representation learning (GRL). Our task is to transform the behaviors in the topological space into low-dimensional representations in a metric vector space while preserving the intra-wise relationships between attributes and inter-wise associations between behaviors.
- Label information learning (LIL). Our task is to project original behaviors into a discriminative vector space, where there is a significant deviation between different behaviors with the help of label information of two granularities.

In this way, the learned attribute representations reflect the data distribution and make it easier to distinguish behaviors. Meanwhile, a low-dimensional space enables efficient computing for subsequent tasks. Note that our transformation network is different from traditional network representation learning algorithms which are used to learn the deep structure information and the label information of nodes or edges by compressing the original input graph.

The core differences are in two aspects: the learning objectives and the object of the labels. Firstly, most network representation learning algorithms are trained for graph reconstruction and inference. They typically obtain low-dimensional representations of nodes, which include both the intrinsic information of the nodes themselves and the structural information within the graph. In contrast, our approach focuses on obtaining more expressive attribute representations. These representations not only capture the structural information of attributes within the graph but also incorporate the relationships between different attributes in the entire behavior space, such as frequent co-occurrence patterns. By leveraging these attribute representations, we can easily identify differences between behaviors, such as deviations in certain attributes or inappropriate co-occurrence of attributes within a behavior. It is advantageous compared to directly detecting representations of behaviors. Secondly, network representation learning algorithms designed for label information primarily focus on nodes or edges within the graph. They typically utilize information propagation and aggregation to directly obtain more discriminative node representations. In our work, the nodes in the graph correspond to behavioral attributes. It is important to note that individual behavioral attributes are neutral and do not inherently have deceptive or nondeceptive characteristics. Only when multiple behavioral attributes are combined to form a behavior (which can be viewed as a subgraph in the graph) do we have corresponding label information. We introduce label information at the subgraph level not for direct node-level classification but rather to transmit information about common deceptive or nondeceptive behaviors to the representations of behavioral attributes.

We introduce the graph representation learning model $GRL_\phi(G)$ to obtain a mapping $\phi(\cdot) : f_i \mapsto \vec{f_i}$ from the attribute f_i in V to its corresponding representation $\vec{f_i}$ in the latent vector space. Specially, it utilizes a network representation learning algorithm to extract deeper information from the behavior graph G. This algorithm generates low-dimensional representations that capture latent patterns, relationships, and structural properties within the graph. We introduce the label information learning model $LIL(\mathscr{B}, \mathscr{Y}^c, \mathscr{Y}^f, \phi)$, which incorporates the label-level information.

3.4.3.1 Structure Information on Behavior Graph

The graph representation learning model $GRL_\phi(G)$ is designed to get the latent representations from the structure of behavior graph. Our $GRL_\phi(G)$ is based on network representation learning algorithms, e.g., network embedding and graph neural networks, due to their success in real applications. $GRL_\phi(G)$ is compatible with the learning process of structural information in many existing algorithms, so other advanced algorithms can also be used as replacement modules to upgrade our method. In a graph, the semantic information of a node is influenced by its local neighborhoods. So we focus on the basic operator on edges to learn the associations between each node and its neighbor nodes. We learn the representations from the following recognized criteria: Given an observed edge (v_i, r, v_j) in the behavior graph, $GRL_\phi(G)$ learns how likely there is an edge of type r from node v_i to v_j. The objective function of $GRL_\phi(G)$ can be expressed as

$$\arg\max_{\vec{v},\vec{r}}\, p(\vec{v}, \vec{r}|G), \tag{3.2}$$

where $\vec{v}$ and $\vec{r}$ are the target optimal node representations and edge representations, respectively. By using Bayes' rule, the goal can be further expressed as

$$\arg\max_{\vec{v},\vec{r}}\, p(G|\vec{v}, \vec{r})p(\vec{v}, \vec{r})/p(G) = \arg\max_{\vec{v},\vec{r}}\, p(G|\vec{v}, \vec{r}),$$

where the denominator $p(G)$ can be eliminated in the process of maximization since it is of no involvement to $\vec{v}$ and $\vec{r}$. We assume that there is no prior knowledge $p(\vec{v}, \vec{r})$ about the optimal representations to simplify the analysis. We also introduce a feasible implementation of how to learn the representations from the observations on the behavior graph.

We employ the neural network model to predict the target edge type r between the pair of nodes v_i and v_j and simplify the prediction into whether the predicted type is correct or not. For each observed edge (v_i, r, v_j), we denote $p(v_i, r, v_j)$ as the probability that our neural network model predicts correctly. It is formulated as the following likelihood function:

$$p(G|\vec{v}, \vec{r}) = \prod\nolimits_{(v_i, r, v_j) \in \mathscr{E}} p(v_i, r, v_j) \tag{3.3}$$

For each observed edge (v_t, v_d, r), we generate the unobserved edge (v_i^u, r^u, v_j^u) by randomly replacing one of the three values with other v_i^u, r^u, and v_j^u. Note that v_i^u and v_j^u are randomly selected from $\mathscr{V}$ and r^u is randomly selected from $\mathscr{R}$. Then, we filter out the "bad" unobserved edges that can be obtained during the generation of observed ones.

Inspired by the score function presented in a representative algorithm, Marine [45], we design the probability function as

$$p(v_i, r, v_j) = \text{Sigmoid}\left(s(v_t, v_d, r) - s\left(v_i^u, r^u, v_j^u\right)\right), \tag{3.4}$$

where the score function is introduced in [45]. It encourages that the gap between $s(v_t, v_d, r)$ and $s(v_i^u, r^u, v_j^u)$ is enlarged. In other words, $p(v_i, r, v_j)$ approaching 1 means that the model correctly predicts r belonging to the pair of v_i^u and v_j^u. Note that the probability function can also refer to other representation learning algorithms. Our $GRL_\phi(G)$ can be compatible with most existing network representation learning algorithms. To facilitate optimization, we minimize the negative logarithm and aim at solving

$$\arg\min\nolimits_{\vec{v}, \vec{r}} L_g = \sum\nolimits_{(v_i, r, v_j) \in \mathscr{E}} - \log p(v_i, r, v_j). \tag{3.5}$$

3.4.3.2 Label Information on Behavioral Data

The label information learning model aims to learn two kinds of grained labels, i.e., coarse labels and fine labels. Based on $\mathscr{Y}^c$ and $\mathscr{Y}^f$, we design a model $LIL(\cdot)$ to map a behavior B in the latent space while increasing the distances between different behaviors (i.e., the relative distance between behaviors with different labels) compared with those in the original space. To achieve these goals, the behaviors in latent space should maintain accurate representations of the behaviors' attributes and make it easier to perform authentication.

Coarse Label: Deceptive or Nondeceptive We first design a naive behavior representation extractor to capture the first-order attribute interactions in the learned metric vector space. Given a behavior $B_i = \{f_1, f_2, \cdots, f_{|B_i|}\}$ and the mapping $\phi(\cdot)$, we formulate the representation of behavior B_i as $\vec{B}_i$:

$$\vec{B}_i = \frac{1}{|B_i|} \sum\nolimits_{i=0}^{|B_i|} \phi(f_i) \tag{3.6}$$

To distinguish behaviors in the representation space, we introduce the cluster centers $\vec{B}^+$ and $\vec{B}^-$ to represent all nondeceptive and deceptive behaviors. The cluster centers $\vec{B}^+$ and $\vec{B}^-$ can be respectively formulated as

$$\vec{B}^+ = \frac{\sum_{B_i \in \mathscr{B}} \max(0, y_i^c) \cdot \vec{B}_i}{\sum_{B_i \in \mathscr{B}} \max(0, y_i^c)},$$
$$\vec{B}^- = \frac{\sum_{B_i \in \mathscr{B}} \min(0, y_i^c) \cdot \vec{B}_i}{\sum_{B_i \in \mathscr{B}} \min(0, y_i^c)}$$

Based on two cluster centers $\vec{B}^+$ and $\vec{B}^-$, we calculate their distance by using the euclidean distance for distinguishing behaviors with different coarse labels. We get the coarse label loss on behavioral data as

$$L_c = \left(\sum\nolimits_{B_i \in \mathscr{B}} |y_i^c|\right) \cdot ||\vec{B}^+ - \vec{B}^-||. \tag{3.7}$$

Fine Label: Behavior Type Inspired by metric learning [46], we train our transformation network by assuming that behaviors with the same type have a smaller distance than those with different types. It makes the learned behavior representations with different types more separable. Specifically, we extract s other behaviors for each behavior B to form s behavior pairs. Let $\mathscr{B}^p$ be a set of behavior pairs; $(B_i, B_j, y_i^f, y_j^f) \in \mathscr{B}^p$ represents a behavior pair for behavior B_i, where $s \cdot |\mathscr{B}| = |\mathscr{B}^p|$; and B_j is the selected behavior to be compared with B_i. Note that B_j is chosen only when $y_j^f > 0$.

Given a pair of behaviors B_i and B_j and their labels y_i^f and y_j^f, we need to ensure that the distance between $\vec{B}_i$ and $\vec{B}_j$ increases if $y_i^f \neq y_j^f$. Hence, the loss can be defined as

$$\begin{aligned} L_f(B_i, B_j, y_i^f, y_j^f) &= \left(1 - \left(y_i^f - y_j^f\right)^0\right) \cdot dis_{ij}^2 \\ &+ \left(y_i^f - y_j^f\right)^0 \cdot \max(0, dis_{max} - dis_{ij})^2, \end{aligned} \tag{3.8}$$

where dis_{ij} is the distance between two behaviors B_i and B_j. The calculation of dis_{ij} is shown below:

$$dis_{ij} = \sum_{i=0}^{|B_i|} ||\phi(f_i) - \phi(f_j)|| \tag{3.9}$$

Note that dis_{max} is a hyperparameter to control the maximum distance between two behaviors with different fine labels, such that dissimilar pairs contribute to the loss function only when their distance is within dis_{max}. Finally, we integrate the multiple behavior pairs into the loss function defined as follows:

$$L_f = \sum_{(B_i, B_j, y_i^f, y_j^f) \in \mathscr{B}P} \frac{\min(1, y_i^f)}{s \cdot (s-1)} \cdot L_f\left(B_i, B_j, y_i^f, y_j^f\right), \tag{3.10}$$

where $\min(1, \cdot)$ ensures that unlabeled data will not influence the loss function.

3.4.3.3 Information Fusion

We have the objective function L for our transformation network as follows:

$$L = \alpha \cdot L_g + L_f + L_c, \tag{3.11}$$

where L has three components: the graph structure loss (L_g) on behavior graph and two label aware losses (L_c and L_f) on behavioral data. Hyperparameter α, $\{\alpha \in \mathbb{R} : 0 \leq \alpha \leq 1\}$, controls the importance of graph structure information during learning the latent representations. Setting it to a high value (close to one) ensures that the model can learn useful representations with little or no labeled behaviors. This strategy encourages the exploration and exploitation of graph information. On the other hand, if α is low (close to 0), it would lead that the graph structure is only used as a way to transmit messages to help identify different behaviors. Finally, we parameterize the final objective function L for efficient representations as $\phi(\cdot)^* = \arg\min(L)$, where $\phi^*(\cdot)$ refers to the mapping function with optimal parameters of our transformation network.

3.4.4 Deceptive Behavior Detection

The representation of behavioral attributes is beneficial for downstream behavior detection tasks, but it is difficult to directly deploy classification models on behavioral attributes to detect deceptive behavior. For this purpose, we design customized features that we compute based on these representation and are not preexisting or standard features. The customized features are designed with the following considerations: First, we are committed to obtaining discriminative features from multiple attributes of behavior in the way of less information loss. We utilize the learned representations and the local structure of attributes, simultaneously. Second, the high-order attribute interactions are useful and have achieved success in many fields. Third, more topological information should be introduced to observe behaviors from different perspectives. Based on the mapping function $\phi^*(\cdot)$, we can transform the behavior B into $\vec{B}^*$ (i.e., the matrix composed of attribute representations):

$$\vec{B}^* = [\vec{f}_1, \vec{f}_2, \cdots, \vec{f}_{|B|}] = [\phi^*(f_1), \phi^*(f_2), \cdots, \phi^*(f_{|B|})]$$

3.4.4.1 Positional Features of Behavior

The first-order attribute interactions (as shown in Eq. (3.6)) are helpful for behavior modeling. In this case, a behavior can be modeled as a sphere composed of its attribute representations in a low-dimensional space, where all representations are located on the surface of the sphere and have the same distance from the kernel. In practical applications, deceptive behaviors often deviate from the patterns of nondeceptive behaviors in special attributes. Obviously, using an average value of the attribute representations to represent behavior is not the best way to model behaviors. A good behavior modeling method should be able to sensitively capture the deviation of attributes in detecting deceptive behavior.

Inspired by object collision detection in the field of computational geometry, we regard the behavior composed of a series of attribute representations as an object composed of multiple vertices in space. Similarly, we introduce a minimum bounding volume (MBV) to judge the intimacy of two behaviors in space. Furthermore, to judge nondeceptive or deceptive behavior, we make the judgment by the distance between its corresponding MBV and others. In our work, we extend the minimum bounding circle algorithm for two dimensions to the dimension of attribute representation to obtain the center and radius (i.e., our customized features) of the bounding sphere in the learned representation space. Given $\vec{B}^*$, we get $\vec{C}, \tau = \varphi(\vec{B}^*)$, where $\varphi(\cdot)$ is the algorithm that computes the minimum bounding volume (MBV), $\vec{C}$ is the center of MBV, and τ is the radius of MBV. Specifically, we use Welzl's algorithm [47] as the deployment of $\varphi(\cdot)$.

3.4.4.2 Internal Relationship Features of Behavior

Deceptive behaviors usually deviate from nondeceptive behaviors on the value of attributes, so the association of any two attributes in a behavior can sensitively identify behavioral deviations. The similarity between two learned attribute representations is one of the good means to evaluate the degree of the association. As shown in Sect. 3.4.2, the behavior subgraph provides us with the attribute pairs that we should pay attention to. One factor that cannot be neglected is that the behavior subgraph provides different graph structures for different behaviors, resulting in a different number of newly constructed features. Structured features are also what we have to consider when applying our idea in more application scenarios. So we calculate the similarities of any two attribute representations as new features based on $\vec{B}^*$. Specifically, we can get $|B| \cdot (|B| - 1)/2$ similarities as the constructed features to reflect the attribute associations. The similarity is formalized as follows: For any two representations $\vec{f_i} \in \vec{B}^*$ and $\vec{f_j} \in \vec{B}^*$, we get $sim(\vec{f_i}, \vec{f_j})$ by using the Cosine similarity [48].

3.4.4.3 Topological Evolution Features of Behavior

The above features focus on exploiting zero-dimensional topology, i.e., the associations among attributes and their local context. Inspired by the persistent homology that is often used to analyze the topological properties of data, we also hope that our customized features are capable of capturing multi-scale structural information in a robust way. In algebraic topology, the persistent diagram can be obtained from the dataset after the persistent homology. For different datasets, even if the number of data samples is the same, the generated persistent diagram may still be different. This uncertainty makes it untenable to learn directly from the persistent diagram as a feature completely. To this end, we turn to vectorization methods to convert persistent diagrams into vectors for machine learning. We adopt the persistent entropy to obtain our new features that incorporate multi-scale topological information into deceptive behavior detection.

3.4.4.4 Detecting Deceptive Behavior

We simplify the compliance authentication task to a binary classification problem to find deceptive behavior from the nondeceptive ones. Specifically, we built the MORE model on collected behavioral data to learn representations. From these representations, we generate customized features and then train a classifier to discriminate behavior. For each incoming behavior requiring detection, we get the representation of each behavioral attribute and calculate customized features. These features are then fed into the pre-trained classifier to obtain predictions corresponding to the behavior.

3.5 Evaluation of MORE

3.5.1 Experimental Setup

3.5.1.1 Security Application

We conduct several experiments on four representative security datasets, i.e., Banking,[3] DNS,[4] CERT,[5] and Darknet.[6] The statistics information of used datasets are listed in Table 3.1. Next, we provide a detailed description of the datasets along with the preprocessing steps undertaken for each dataset.

Fraudulent Transaction Detection on Banking We deploy the experiment and test our method in collaboration with a prestigious bank in China. The online payment dataset had been collected and desensitized by the bank. Before data collection, all users were informed by statement subject to local privacy policies. In this work, we solely focus on the association of behavioral attributes while building the behavior graph and did not require access to the original data (including the sensitive data). So we rely on the data provider to provide us with the corresponding pre-trained word vectors on behavioral attributes. The pre-trained word vectors capture semantic relationships and contextual information between behavioral attributes and generally do not hold private or sensitive information. Then, in the deployment of our transformation network, we use a set of hashed attributes as nodes without touching actual attribute values in the behavior graph. The dataset contains six different behavior types (fine labels), including two nondeceptive classes and four deceptive classes. All the transactions are labeled by the bank manually. For deceptive behaviors, data is labeled by fraud tricks (e.g., Trojan virus, telecommunications, phishing websites, and others).

Table 3.1 The statistics information of four datasets. N# means number, and R# represents ratio

Dataset	N# of behaviors	N# of deceptive behaviors	R# of deceptive behaviors	N# of deceptive classes	N# of behavioral attributes
Banking	1,028,437	24,898	2.42%	4	8
DNS	50,159	25,910	51.65%	3	13
CERT	354,667	5,087	1.43%	5	8
Darknet	141,530	24,311	17.18%	8	9

[3] *Banking*:The dataset is not publicly available due to the restrictions of privacy protection and business secrets. Data can only be used for research after desensitization. Prior to accessing the dataset, all researchers must agree to the bank's Developer Agreement and Policy.

[4] *DNS*: https://www.unb.ca/cic/datasets/dns-2021.html.

[5] *CERT*: https://kilthub.cmu.edu/articles/dataset/Insider_Threat_Test_Dataset/12841247.

[6] *Darknet*: https://www.unb.ca/cic/datasets/darknet2020.html.

Table 3.2 Selected attributes in Banking dataset

	Attribute	Description
Banking	faccount_number	Each account_number represents a user's account
	merchant_number	Each merchant_number represents a merchant in a B2C transaction
	place_number	Each place_number represents an issuing area of banking cards used for transactions
	time	The exact time when the transaction occurred
	amount	The amount of money transferred to the merchant in a transaction
	ip	Whether a commonly used ip or not in a transaction
	last_result	Judgment of the last transaction in the relevant account_number
	type	Each type represents a transaction of different type

We adopt payment transactions from June 2017 as the experimental dataset. The dataset contains 1,028,437 samples, including 1 nondeceptive behavior class of 1,003,539 samples and 1 deceptive behavior class of 24,898 samples. In the original dataset, each sample representing a payment transaction is characterized by 64 behavioral attributes. We chose eight behavioral attributes (as shown in Table 3.2) in all to represent a behavior instead of adopting them completely. Both transaction amount and time are typically continuous values. So we discretize the transaction time into four eight (4×2) intervals. We set four small intervals as follows: $[0, 3)$, $[6, 11)$, $[15, 24)$, and $[3, 6] \cup [11, 15]$. Meanwhile, we divide the time by distinguishing whether the transaction time falls on a working day. Regarding the transaction amount, we discretize it into four different values based on the distribution in the data. The intervals used for discretization are $[0, 60)$, $[60, 3000)$, $[300, 3600)$, and above 3600.

Deceptive Attack Detection on DNS We use an open released dataset *DNS* where threat actors usually abuse the DNS (domain name system) to lure users to be victims of malicious domain hosting [27]. All these samples were processed from 1 million benign domains and 51,453 known deceptive domains in public datasets. The deceptive samples span three classes: spam, phishing, and malware. We choose the new version of this dataset (to be released in 2022). It consists of 24,249 nondeceptive samples and 25,910 deceptive samples processed from a million nondeceptive and known malicious domains in publicly available datasets. Each sample consists of 32 clearly defined behavioral attributes, including lexical-based, DNS statistical-based, and third-party-based attributes, where the attributes are extracted from the captured DNS PCAP files. In our work, we select the top 13 influential attributes (as shown in Table 3.3) through evaluating them using information gain.

Threatening Action Detection on CERT In the insider threat dataset *CERT*, all the system behaviors (including nondeceptive background data and data from

Table 3.3 Selected attributes in DNS dataset

	Attribute	Description
DNS	subdomain	Has subdomain or not
	tld	Top-level domain
	sld	Second-level domain
	len	Length of domain and subdomain
	numeric-percentage	Counts the number of digits in domain and subdomain
	entropy	Entropy of letter distribution
	longest-word	Longest meaningful word in SLD
	typos	Typosquatting
	obfuscate-at-sign	Max value for URL obfuscation
	Domain-Name	Name of the domain
	Domain-Age	The age of a domain
	State	The state the main branch is
	Name-Server-Count	The total number of name servers linked to the domain

synthetic deceptive actors) are recorded in five files separately that contain log-in operations, removable storage device usage, file operations, network operations, and email traffics [49]. We use the r4.2 version of this dataset to conduct our experiments. Different operations are recorded in different files, so we label the classes of behaviors according to the file division. The CERT dataset used in our study comprises 32,770,227 user operations from 1,000 users, covering 3 attack scenarios, including various insider threats. The original dataset covers a time span from January 2010 to June 2011. In this study, we focused on the behavioral data from the second half of 2010. It is within this time range that the majority of deceptive activities occurred. Furthermore, the number of nondeceptive behaviors is considerably lower than the number of deceptive behaviors. To alleviate this class imbalance, we performed downsampling on the nondeceptive behavioral data, retaining only 0.05% of the nondeceptive behaviors. We record the behavior as a set of behavioral attributes composed of user, operation host, activity status, and occurring time. Each sample is represented as a vector of eight behavioral attributes (containing five different granularity of time), indicating the user behavior of corresponding system operations (as shown in Table 3.4).

Suspicious Traffic Detection on Darknet Any traffic communications from dark space are contemplated to be unsought and are characteristically treated as probes, where the host is not speculated to interact with other computers. We use a darknet traffic dataset generated by the Canadian Institute for Cybersecurity [25]. The deceptive darknet traffic constitutes audio stream, browsing, chat, email, P2P, transfer, video stream, and VOIP which are generated by the special applications. The dataset contains nine different classes, including one nondeceptive class of 117,219 samples and eight deceptive classes of 24,311 samples. Each sample is represented as a vector of 83 behavioral attributes, which are generated by deploying ISCXFlowMeter to read and process PCAP files. To facilitate the demonstration,

Table 3.4 Selected attributes in CERT dataset

	Attribute	Description
CERT	user	User's ID
	pc	The ID of the terminal device used by the user
	state	The activity type of the behavioral operation
	month	The month extracted from the date field of the behavior operation
	weekday	The weekday extracted from the date field of the behavior operation
	day	The day extracted from the date field of the behavior operation
	hour	The hour extracted from the date field of the behavior operation
	second	The second extracted from the date field of the behavior operation

Table 3.5 Selected attributes in Darknet dataset

	Attribute	Description
Darknet	Idle-Max	The maximum amount of time a network flow was idle before becoming active
	Fwd-Seg-Size-Min	The minimum forward segmentation size of a data packet
	Bwd-Packet-Length-Min	The minimum length of a network packet during transmission in the backward (receiving) direction
	Protocol	The type of transmission protocol used by network packets at the transport layer
	Idel-Mean	The mean amount of time a network flow was idle before becoming active
	FWD-Init-Win-Bytes	The initial sliding window size value when the connection is established in the forward direction
	FIN-Flag-Count	The number of segments with the end flag FIN set in a network flow or connection
	Sub-flow-Bwd-Bytes	The total amount of traffic size of each subtraffic in a network flow during the receiving process
	Bwd-Packet-Length-Max	The maximum length of a packet during transmission in the backward (receiving) direction in a network flow or connection

we indicate the traffic behaviors based on nine selected attributes (as shown in Table 3.5) that are selected as the most important attributes in a simple tree-based model.

3.5.1.2 Evaluation Metrics

We use the standardized partial AUC ($\text{SPAUC}_{\text{FPR}\leq\text{maxfpr}}$) as the primary metric [35]. It denotes the area of the head of the ROC curve when FPR $\leq$ maxfpr. The $\text{SPAUC}_{\text{FPR}\leq\text{maxfpr}}$ is calculated as follows,

$$\text{SPAUC}_{\text{FPR}\leq\text{maxfpr}} = \frac{1}{2}\left(1 + \frac{\text{AUC}_{\text{FPR}\leq\text{maxfpr}} - \text{minarea}}{\text{maxarea} - \text{minarea}}\right),$$

where maxarea = maxfpr and minarea = $(\text{maxfpr}^2)/2$. To express concisely, we use AUC to refer to $\text{SPAUC}_{\text{FPR}\leq\text{maxfpr}}$ for all experiments. In different real-world scenarios, we set different maxfpr according to the de facto standards. In this work, we set maxfpr to 0.001, 0.005, 0.01, and 0.05 for the Banking, DNS, CERT, and Darknet datasets, respectively. In addition, we also report precision, recall (TPR), disturbance (FPR), and F1 score to evaluate our model comprehensively.

3.5.1.3 Implementation

The experiments are conducted on a local PC device based on the Windows 10 Pro operating system. The device is equipped with an Intel(R) Core(TM) i7-9700K CPU (8 cores, 3.6 GHz), 64 GB of RAM, and an NVIDIA GeForce RTX 2080 Ti (11 GB) graphics card. All experiments are implemented using the Python 3.7 programming language. We rely on the scikit-learn 1.0 and PyTorch 1.11 frameworks for machine learning and neural network-related code implementation. Additionally, we utilize version 0.8 of DGL, a library specialized in graph deep learning, which provides efficient graph representation and manipulation capabilities, for the graph learning component. For data handling and preprocessing tasks, we primarily rely on Pandas 1.3, a powerful library for data processing and analysis, which facilitates the loading, cleaning, and transformation of the experimental data. If not otherwise stated, we split all the dataset into training set, unlabeled set, and testing set with a ratio of 20 : 40 : 40. In the training set, we pick 20% of the samples as the validation set. We report the average results after ten independent runs on the testing set.

3.5.2 *Comparison with Baselines*

Experimental Design We mainly compare our method with several representative supervised methods (logistic regression (LR), random forest (RF), CatB [50], and TabNet [31]), two popular semi-supervised methods (label propagation (LP) and DevNet [51]), and the graph-based methods (BIG [28], HGT [52]). In this experiment, all methods use a combination of selected behavioral attributes as input. Our method also takes the selected attribute as input and then calculates customized features to ultimately evaluate the performance. We utilize several baseline methods for the above four datasets. The three classic baselines (LR, RF, and LP) have been highly cited and widely used in classification tasks [53, 54]. We use the implementation of LR, RF, and LP available at the scikit-learn Python package. CatB[7] is a kind of boosting family algorithm [55], which is widely used in the industry and can rival advanced machine learning algorithms [56]. DevNet[8] is the

[7] *CatB*: https://catboost.ai.

[8] *DevNet*: https://github.com/GuansongPang/deviation-network.

advanced anomaly detection approach, which synthesizes neural networks, gaussian prior, and z-score-based deviation loss to perform data-efficient and achieves significantly better anomaly detection based on a few labeled anomalies. TabNet[9] is an attentive transformer neural network proposed by Google Cloud AI. It is designed for tabular data and has proven to be superior to or equal to other advanced models for classification in different fields. BIG[10] is a graph-based behavioral modeling solution by mining multiple-level associations among behaviors and is validated by many representative security issues. HGT[11], as an advanced method based on graph neural networks, maintains the dedicated representations for different types of nodes and edges in the graph structure and performs well in node classification tasks. Based on the original behavioral data, we design two graph structures and input them into HGT to implement the behavior classification task, which is recorded as HGT-hn and HGT-hn, respectively. For HGT-hn, we build a heterogeneous graph that contains multiple nodes of behavior and behavioral attributes. For each pair of appearing behavior (ones attribute), we defined it as the edge. The graph for HGT-hn contains an intuitive association between behaviors (i.e., behaviors with the same attributes have similar contexts in the graph structure). For HGT-he, we build a graph that contains only one kind of node (behavior) and multiple edge types. For each behavior, we define it as a node. Then, five edges are generated for each attribute's category where the types of edges depend on the categories of attributes. The edge means that two connected behaviors have the same attribute. In addition, we take the representation of behavior as the initial input of the node embedding in HGT-he. In HGT-hn, the attributes are regarded as nodes in the graph, so we only randomly select an initial embedding for each node as input.

Table 3.6 summarizes the performance of all methods.

Ours vs. Supervised Methods We find that the AUC of ours is consistently higher than all the supervised methods on all datasets. It validates our hypothesis that the representation of behavior is limited in modeling behavior patterns and their performance is lower than our methods even when the model and data distribution are well matched. Note that the performance of LR seems to be a little too tragic. The phenomenon of such a simple basic model is also close to reality when deliberate attackers are good at camouflaging their deceptive intents. Here, we focus on the performance under low FPR, which is difficult to distinguish deceptive behaviors in this case. Ours has high TPR (about 5% improvements on DNS, 45% on CERT, and 7% on Darknet), indicating it can capture more deceptive behaviors under the same restrictions.

Ours vs. Semi-Supervised Methods Semi-supervised methods do bring performance gains by learning information from unlabeled behaviors, but the gains are limited. Our method can also add unlabeled behaviors to the representation learning

---

[9] *TabNet*: https://github.com/google-research/google-research/tree/master/tabnet.

[10] *BIG*: https://github.com/SlothfulZhu/Wrongdoing-Monitor.

[11] *HGT*: https://github.com/dmlc/dgl/tree/master/examples/pytorch/hgt.

Table 3.6 Performance comparison of different approaches

Dataset	Banking			DNS			CERT			Darknet		
	maxfpr=0.001			maxfpr=0.005			maxfpr=0.01			maxfpr=0.05		
Metric	AUC	TPR	Prec	AUC	TPR	Prec	AUC	TPR	Prec	AUC	TPR	Prec
LR	0.52	0.05	0.52	0.53	0.07	0.94	0.52	0.06	0.07	0.50	0.09	0.26
RF	0.89	0.88	0.96	0.68	0.41	0.99	0.66	0.41	0.38	0.80	0.84	0.78
CatB	0.88	**0.89**	**0.96**	0.96	0.93	1.00	0.70	0.54	0.44	0.83	0.86	0.79
TabNet	0.79	0.87	0.96	0.81	0.73	0.99	0.55	0.14	0.16	0.72	0.76	0.75
LP	0.88	0.82	0.95	0.81	0.63	**1.00**	0.72	0.60	0.46	0.87	0.92	0.79
DevNet	0.51	0.04	0.46	0.63	0.08	0.94	0.53	0.11	0.01	0.64	0.58	0.68
BIG	0.88	0.88	0.96	0.96	0.96	1.00	0.82	0.89	**0.88**	0.87	0.92	0.79
HGT-hn	0.89	0.88	0.96	0.95	**0.98**	1.00	0.93	0.92	0.57	0.84	0.89	0.79
HGT-he	0.50	0.00	0.07	0.58	0.22	0.98	0.51	0.04	0.06	0.54	0.18	0.42
Ours	**0.91**	0.88	0.95	**0.99**	0.98	1.00	**0.94**	**0.96**	0.58	**0.88**	**0.93**	**0.79**

process, so it maintains similar performance gains. DevNet learns data distribution for identifying anomalous samples from normal ones. But the abnormal behaviors are not exactly consistent with deceptive behaviors, so its performance is poor at low disturbance rates. At the same time, we also observe that DevNet has a significant improvement compared with LR through the customized learning of abnormal samples. The use of unlabeled information helps improve authentication in the face of deceptive behaviors. Our method can learn from unlabeled data, which we have verified in the subsequent experiments (Sect. 3.5.6).

Ours vs. Graph-Based Methods Existing graph-based methods can model the associations between behaviors by introducing graph structure. As HGT-hn and BIG show, they are obviously effective and achieved suboptimal performance. On the contrary, the improvement made by HGT-he is very limited and cannot break through the limits of the traditional representation learning method that realizes information propagation and aggregation based on the representation of behaviors. In this case, the association between behaviors is added to facilitate the learning of context information of similar behaviors. We find that there is still a gap between HGT-hn and our method, especially on the CERT and Darknet datasets. This proves that it is necessary to understand the structure of behavior in an appropriate way.

3.5.3 Performance in Identifying Behavior Types

Experimental Design Assuming that we have detected various types of deceptive behaviors across different scenarios, it is evident that identifying the specific behavior types would greatly assist defenders in analyzing how to prevent such behaviors. This experiment focuses on the performance of abstracting behaviors (i.e., correctly identifying fine labels in multi-classification tasks). For each dataset,

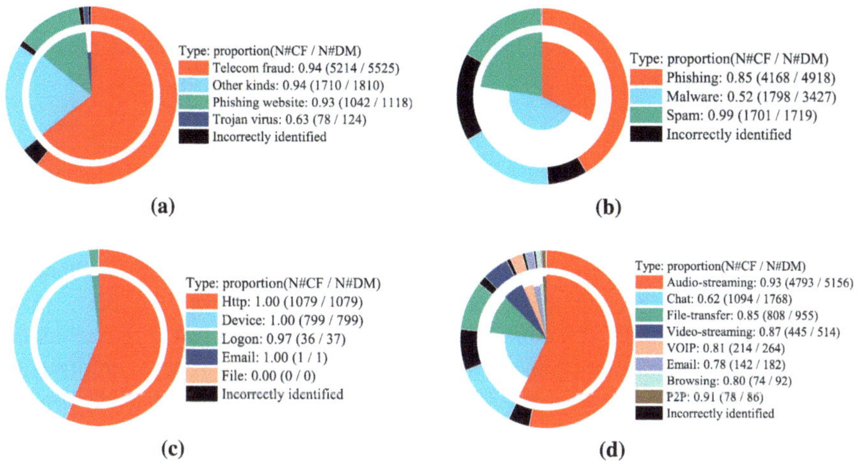

Fig. 3.4 MORE's detection results on fine labels. (**a**) Banking, TPR: 0.88, N#UDM: 1192. (**b**) DNS, TPR: 0.98, N#UDM: 256. (**c**) CERT, TPR: 0.96, N#UDM: 73. (**d**) Darknet, TPR: 0.93, N#UDM: 690

we capture the deceptive behaviors according to the deployment in Sect. 3.5.2. Meanwhile, we employ a classifier to implement multi-classification tasks of fine labels and give each behavior a predicted behavior type. We only retain the true positive behaviors (i.e., captured deceptive behaviors) in the task of identifying wrongdoings to verify the accuracy of the behavior abstraction. Figure 3.4 summarizes the behavior abstraction accuracy of our method on four datasets. The metric is the number of deceptive behaviors that are correctly identified on fine labels (denoted as N#CF)/the number of total detected deceptive behaviors (denoted as N#DM). Different colors represent different types of behaviors, while black specifically indicates that deceptive behaviors are incorrectly identified on fine labels. Furthermore, we annotate the capture rate (TPR) for deceptive behaviors and the number of undetected deceptive behaviors (N#UDM) for each dataset.

Findings Although MORE does not correctly identify the type of all deceptive behaviors, it achieves good performance (more than 80% accuracy) on abstracting behaviors in most cases. Our method shows its generic to multiple distinct behaviors. For instance, in Banking dataset, our method can correctly identify more than 93% of deceptive behaviors in the face of telecommunications fraud, phishing website fraud, and other frauds with the unknown behavior type. When faced with the fraud of implanting Trojan virus into mobile phones, due to its small proportion, the capture rate of our model drops to only 63%. Note that capture here refers to our model's ability to abstract behaviors (i.e., correctly identify the type of deceptive behavior), not to omit deceptive behaviors. In CERT dataset, we almost correctly identify all classes of deceptive behaviors. An unexpected zero in Fig. 3.4c is attributed to the number of deceptive behaviors about file operations being too small (the percentage of single digits in the test dataset), which makes the index easy to fluctuate and has a limited impact on the overall performance. When faced with eight classes of deceptive behaviors in dark network traffics, our method is stable in most classes.

3.5.4 *Visualization of Behaviors*

Experimental Design To intuitively reflect the effectiveness of our method, we use t-SNE [57] to project the behaviors into a 2D plane, which allows us to visualize the behavior distribution. T-SNE can project high-dimensional embedding space into a two-dimensional space while largely preserving structural information among embeddings. Each behavior is represented as the embedding consisting of the original attributes or customized features, where the spatial distances between embeddings express their distribution. To manage the complexity, we randomly sample 2,000 behaviors (1,000 deceptive and 1,000 nondeceptive behaviors) from the dataset for visualization purposes. By using this subset of behaviors, we obtain a scatterplot of 2,000 points, wherein each point represents one of these behaviors. Meanwhile, we use the marker of the point to indicate its behavior type. Figure 3.5 shows the 2D visualization of behavior distribution.

Findings Points are distributed in different clusters, suggesting that behaviors are indeed grouped and driven by some potential patterns. Based on the original attributes, the nondeceptive and deceptive behaviors are mixed, as shown in Fig. 3.5.

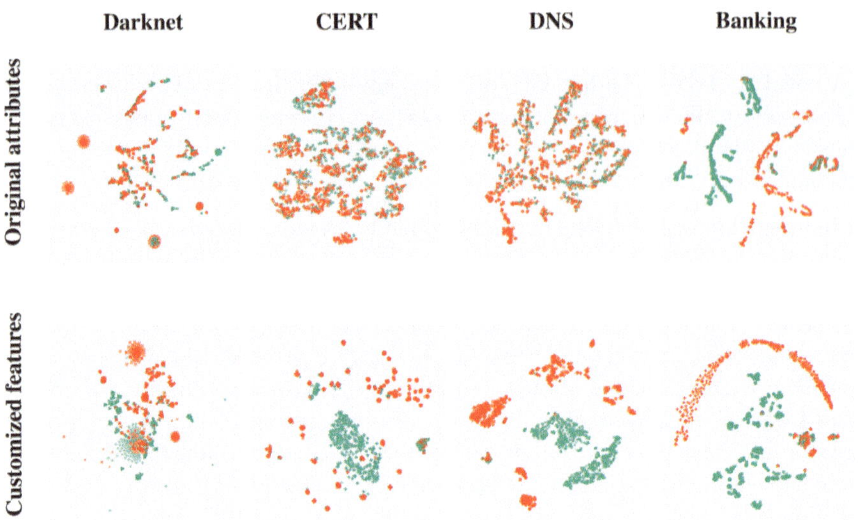

Fig. 3.5 Visualization of the behavior distribution on 1,000 deceptive behaviors (in red) and 1,000 nondeceptive behaviors (in green)

By modeling the structure of behavior, the nondeceptive and deceptive behaviors can be roughly separated apart from each other. As shown in Fig. 3.5, we get some behavior groups that have been separated in the version of customized features. In addition, in the version of original attributes, there is no clear margin between different clusters, and deceptive behaviors with different types are mixed together. In most cases, our method can achieve large margins among different clusters. We observe that our method, compared to the version based on original attributes, not only maintains the classification of behaviors as deceptive or nondeceptive but also further divides the behaviors based on behavior types. The deceptive behaviors are divided into multiple different clusters, which will help us further identify the specific behavior types. Similarly, there are more small clusters that correspond to different nondeceptive behavior patterns. The above results indicate that by modeling the structure of behavior our method can be applied to detect deceptive behavior with good performance.

3.5.5 Ablation Study

3.5.5.1 Sub-Component Evaluation

We conduct a series of experiments to assess model's effectiveness by comparing the performance of different components without the model.

Experimental Design We design experiments from three aspects: loss function, behavior subgraph generation strategy, and similarity measurement strategy. Regarding the loss function, we remove the fine label, coarse label, and graph structure information, separately. For the behavior subgraph generation strategy, we use randomly or fully connected strategies. We also conduct experiments using the Cosine similarity in behavior subgraph and the euclidean metric in generating customized features.

Findings The experimental results are shown in Table 3.7. In most cases, our method achieves the best performance. We observe that the fine label plays a more crucial role in the Banking and DNS datasets, as removing these labels leads to a significant decrease in performance. However, the impact is less pronounced in the CERT and Darknet datasets. This phenomenon arises from the specific deceptive behavior attack patterns that are reflected in the former datasets, which are strongly correlated with the coarse label, thus significantly improving the ability to detect deceptive behaviors. Regarding the behavior subgraph generation strategy, we observe that constructing the subgraph using a fully connected approach leads to a more noticeable performance decrease. This confirms that excessive connections can introduce irrelevant information, highlighting the importance of leveraging the associations between behaviors for detecting deceptive behaviors. When constructing the behavior subgraph, using the Cosine similarity proves to be a feasible strategy, with only a limited performance decrease. On the other hand, in

Table 3.7 Variant comparison

Dataset	Banking	DNS	CERT	Darknet
Without fine label	0.8826	0.8540	0.9650	0.8725
Without coarse label	0.8980	0.9779	0.8355	0.8784
Without graph structure information	0.9013	0.9805	0.9602	0.8790
Randomly link in the behavior subgraph	0.9057	0.9831	0.9229	0.8761
Fully link in the behavior subgraph	0.9073	0.9750	0.8793	0.8762
Cosine in generating behavior subgraph	0.9058	0.9715	0.8883	0.8765
Euclidean metric in customized features	0.9040	0.9608	0.8391	0.8741
Ours	0.9113	0.9867	0.9365	0.8789

the process of generating customized features, Cosine similarity plays a significant role in identifying behavioral differences, outperforming the variant that employs the euclidean metric.

3.5.5.2 Sub-Feature Evaluation

We construct the customized features to mimic the scenarios where a certain number of features are missing to demonstrate how each component of them contributes to the overall performance.

Experimental Design For each feature discussed in Sect. 3.4.4, we exclusively rely on a single type of feature for our experiments. We label these features as "Miniball," "Cosine," and "Homology" to distinguish their respective implementations and evaluations. "Miniball" captures the positional information of behavior in the learned representation space. It extends the minimum bounding volume algorithm to higher dimensions, obtaining the center and radius of the bounding volume as features. "Cosine" utilizes cosine similarity between behavioral attributes as a feature to assess the deviation of behavior from attributes. "Homology" utilizes persistent entropy in topology to calculate the behavioral attribute representation, obtaining a feature vector that reflects multi-scale topological information of behavior. In addition, we provide the version based on the original attributes for comparison, denoted as "Ori," and the version based on all customized features for comparison, denoted as "All." We show the partial ROC curves in Fig. 3.6.

Findings "All" (i.e., the version using "Miniball," "Cosine," and "Homology," simultaneously) consistently outperforms other methods on DNS, CERT, and Darknet. This is a good demonstration that our features based on the learned representations can improve performance in compliance authentication. On the Banking dataset, "Cosine" achieves the best performance, and "All" achieved almost sub-best performance under the low FPR. Specifically, our designed features capture the inconsistency of behaviors based on their high-dimensional structure. The principle is that the value of attributes from different behaviors may share

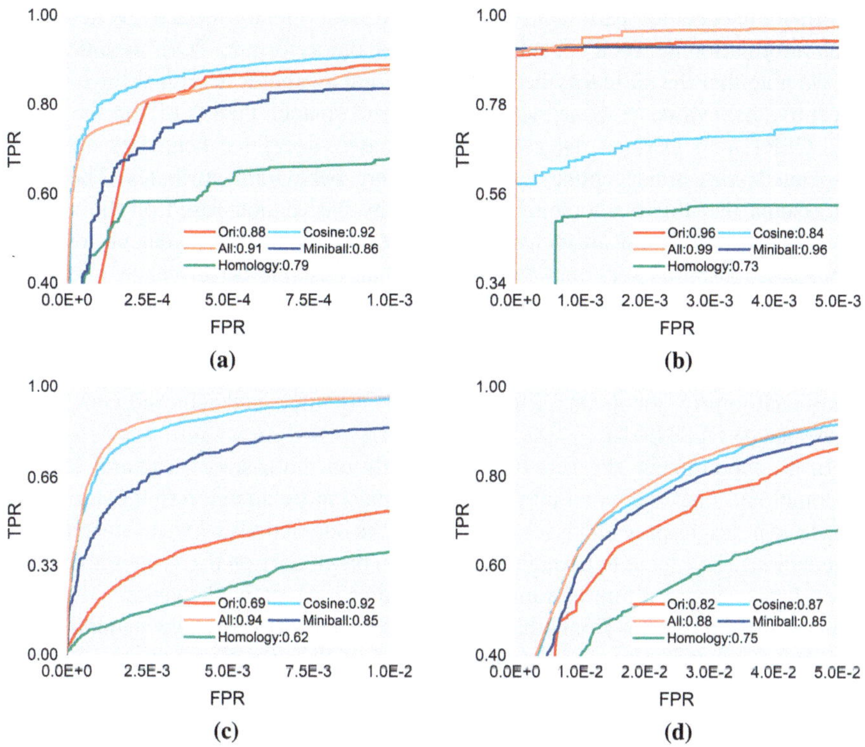

Fig. 3.6 Comparisons among MORE and different customized feature strategies. For each strategy, we report in the legend the SPAUC value obtained in the task of identifying deceptive behavior (e.g., Ori: 0.88). (**a**) Banking. (**b**) DNS. (**c**) CERT. (**d**) Darknet

similar roles and semantics, so behaviors with similar attributes should have a similar structure. For example, when fraudsters cash in, they tend to transfer money by buying high-value goods that match users' preferences. To do this before users notice their accounts have been stolen, fraudsters may browse for only a few minutes before placing expensive purchases. In this case, the fraudster mimics the purchasing preferences of the user, which results in the representation of deceptive behavior that is nearly that of nondeceptive behaviors in the low-dimensional space. Because there is a semantic mismatch between quick purchases and high-value goods for the user, our method can easily detect such deceptive behaviors. Our method models the structure of behavior by learning representations on behavioral attributes, which can capture high-dimensional structural information (e.g., the behavioral similarity that is seemingly mismatched in semantics) that is indeed neglected in low-dimensional space. In the CERT dataset, behavioral attributes are significantly sparse (although there are eight behavioral attributes, five of them are time-related). Thus, the importance of high-dimensional structural information becomes particularly pronounced. Our method effectively leverages such information between behavioral attributes, outperforming "Ori" significantly.

We note that the similarity between behavioral attributes is critical for inferring deceptive behaviors. "Cosine" shows better performance than "Ori" on the Banking, CERT, and Darknet datasets. In most cases, deceptive behaviors are often associated with nondeceptive behaviors in rare behavioral attributes. Therefore, we assume that this phenomenon explains why the version based on the similarity between behavioral attributes can achieve significant improvements over the representation-based versions. "Miniball" has an unstable performance, partly due to its approximation to the representation-based version. Although it is sensitive to capturing the deviation of behavioral attributes, it is still addicted to capturing associations in low-dimensional space. Compared with "Ori," "Miniball" has made significant improvements on CERT and Darknet datasets and achieved comparable performance on DNS.

In all experiments, the version relying only on multi-scale structural features ("Homology") has the worst performance. It pays more attention to the number of rings and holes in space and is less sensitive to the interactions between attribute representations. This leads to a rough classification based only on the structural features of all behaviors, resulting in a low ability to detect deceptive behaviors. However, considering the features from other perspectives, it can provide a new understanding for identifying such seemingly identical behaviors. "Miniball" and "Cosine" ensure that most behaviors can be effectively identified, while "Homology" enhances the recognition of indistinguishable behaviors in our method. Naturally, "All" can obtain the best performance in most cases.

3.5.6 Model Sensitivity to Hyperparameters

Experimental Design To explore the benefits of using (partial) labels, we train MORE with different proportions of labeled behaviors. For each setting, we run our method on randomly sampled behavioral data, where each experimental group has a proportion of deceptive behaviors that is consistent with the original dataset. We report the standardized partial AUC in Fig. 3.7a. The performance of all experiments depended on the detection of coarse-grained labels. In addition, we evaluate how the key parameter of MORE affects the detection performance in Fig. 3.7.

More Labels for Better Performance Figure 3.7a shows the performance of our method under different proportions of labeled behaviors. It shows that the performance on all datasets increases as the proportion increases. This means our method has the capability in dealing with label information during representation learning. Labeled behaviors have a significant impact only when the proportion is small. When we provide only 5% of labeled behaviors for training, our method will degrade with the performance but still have competitive performance in some

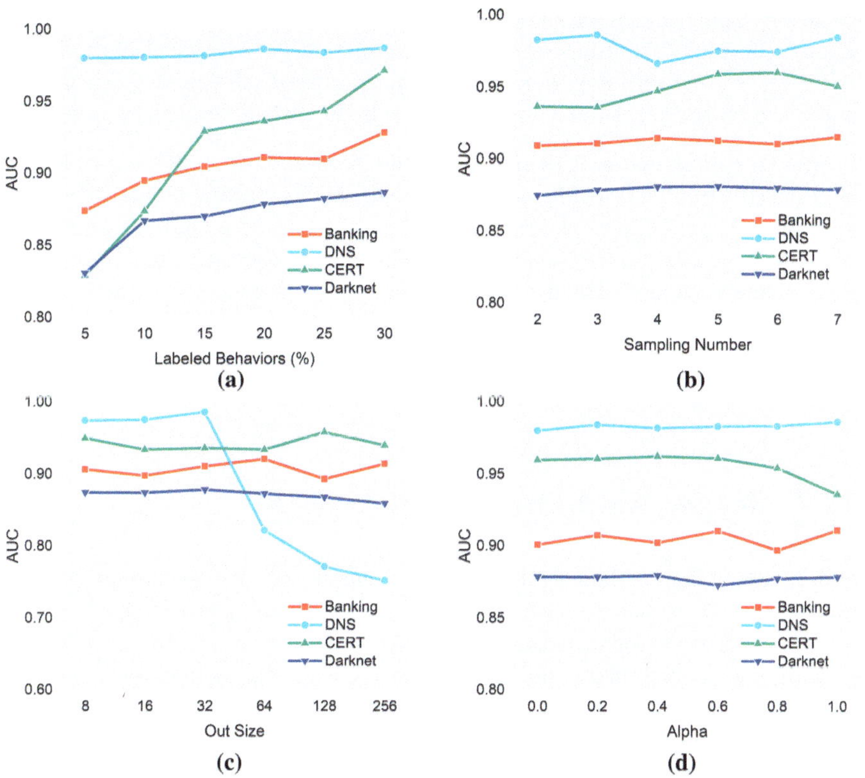

Fig. 3.7 Performance of MORE in four datasets with different amounts of labeled behavioral data

cases. It indicates that label information acts as the enhancement element than the necessary element in our method.

Pair-Wise Metric Learning is Effective Figure 3.7b shows the result of our method with different sampling number s during representation learning. We vary s from 2 to 7 to study the learning ability of pair-wise loss. We can make observations that the performance is respectable even in the face of a small sampled number of behavior pairs. As we add more behavior pairs to our transformation network, the AUC is increasing slightly when s varies from 2 to 7 but only by a small margin. It is sufficient with four or even three behavior pairs to learn the differences in behavior. Continue to increase s; the performance gain is diminishing.

The Representations are Effective Enough Figure 3.7c shows the performance under different dimensions of learned representations. When the dimension set becomes larger, the performance on all datasets does not increase significantly. On the DNS dataset, the performance decreases significantly when it is greater than 32. Recalling the behavior numbers of datasets, DNS only contains about 50,000 behaviors, which is significantly smaller than others. This indicates that a small value is capable of capturing the behavior information, but a large value may lead to noises and cause overfitting. In most cases, an appropriate choice of dimension will improve performance, but the improvement is limited when setting a small value.

Proper Graph Structure is Beneficial We do the analysis to study the contribution of graph structure on detecting deceptive behaviors by adjusting the loss weight on the structure information. The results are shown in Fig. 3.7d. Our method reduces AUC by about 0.01 on the Banking and DNS datasets when structural information is completely omitted. On the contrary, the performance on the CERT dataset declines with more consideration of the structure information. This means that our method is sensitive to structural information and has the potential to further improve performance by carefully designing the graph structure.

3.5.7 Runtime and Memory Cost Breakdown

In this part, taking the Banking dataset as an example, we calculate the computation overhead of our model. According to the experimental setup provided, we test the model's overhead in identifying deceptive behavior tasks.

During the testing phase, the computational cost of our method for identifying the target behaviors primarily depends on the process of generating customized features based on the representations of behavioral attributes. It takes approximately 12 milliseconds to compute customized features for each online payment behavior. The memory cost in this process is approximately 907 MB, mainly due to the storage of learned representations of behavioral attributes. The entire training process of the model consists of data preprocessing, training of the transformation network, generation of customized features, and training of the classification model. As both data preprocessing and the generation of customized features are performed independently for each behavior, multi-threaded synchronous operations can be employed to accelerate the processing speed. The memory cost during the data preprocessing process is approximately 3.4 GB, with an average computation time of 21 milliseconds per behavior for generating behavior subgraph. In our work, the training of the transformation network depends on the size of the behavior graph. Based on 1 million behavior records, it generates the behavior graph with approximately 19.2 million edges, occupying around 2.6 GB of GPU memory resources. The training of the model can be completed in approximately 30 minutes at the fastest. After this step, customized features need to be generated for every behavior in the training set. For a training set of around 200,000 samples, this takes approximately 41 minutes on a single core (optionally reduced to around 6 minutes with multi-threading on our single machine). The training time of the classification model depends on the specific model deployed. In this study, we utilized the LightGBM model that can complete the train in less than 1 minute.

3.6 Conclusion

We revisit the problem of existing security techniques for detecting deceptive behaviors and introduce a feasible paradigm, behavior reconstruction, to design defenses in cyberspace. We propose a novel method, MORE, that encodes behaviors as motifs of behavioral attributes and transforms behavioral attributes into attribute representations. By utilizing the information advantage of defenders, it perceives behaviors from behavioral attributes based on behavioral structure, making it more effective than those based on the representation of behavior. Our work raises the initial understanding of the high-dimensional structure of behaviors and reveals the effectiveness of behavioral structure in detecting deceptive behaviors. The extensive experiments manifest that our method can verify whether user behavior is nondeceptive or deceptive with a small amount of user disturbance. In future research, we aim to explore adaptive evolving detection mechanisms to further address the challenges posed by attackers' evasion tactics.

References

1. J. Zeng, Z.L. Chua, Y. Chen, K. Ji, Z. Liang, J. Mao, in *Proceedings of the NDSS 2021, Virtual Event, February 21–25, 2021* (2021)
2. C. Wang, C. Wang, H. Zhu, J. Cui, IEEE Trans. Dependable Secure Comput. **18**(5), 2122 (2021)
3. X. Ruan, Z. Wu, H. Wang, S. Jajodia, IEEE Trans. Inf. Forensics Secur. **11**(1), 176 (2016)
4. S.M. Milajerdi, B. Eshete, R. Gjomemo, V.N. Venkatakrishnan, in *Proceedings of the CCS 2019, London, UK, November 11–15, 2019* (2019), pp. 1813–1830
5. W. Bialek, Proc. Natl. Acad. Sci. **119**(18), e2021860119 (2022)
6. S. Xi, S. Yang, X. Xiao, Y. Yao, Y. Xiong, F. Xu, H. Wang, P. Gao, Z. Liu, F. Xu, J. Lu, in *Proceedings of the CCS 2019, London, UK, November 11–15, 2019* (2019), pp. 2421–2436
7. Y. Chen, S. Wang, Y. Qin, X. Liao, S. Jana, D.A. Wagner, in *Proceedings of the CCS 2021, Virtual Event, Republic of Korea, November 15–19, 2021* (2021), pp. 477–494
8. L. Tong, B. Li, C. Hajaj, C. Xiao, N. Zhang, Y. Vorobeychik, in *Proceedings of the USENIX 2019, Santa Clara, CA, USA, August 14–16, 2019* (2019), pp. 285–302
9. X. He, Q. Gong, Y. Chen, Y. Zhang, X. Wang, X. Fu, IEEE Trans. Dependable Secure Comput. **18**(5), 2193 (2021)
10. P. Zhang, Z. Sun, S. Kyung, H.W. Behrens, Z.L. Basque, H. Cho, A. Oest, R. Wang, T. Bao, Y. Shoshitaishvili, G. Ahn, A. Doupé, in *Proceedings of the CCS 2022, Los Angeles, CA, USA, November 7–11, 2022* (2022), pp. 3165–3179
11. H. Kwon, A. Mohaisen, J. Woo, Y. Kim, E. Lee, H.K. Kim, IEEE Trans. Inf. Forensics Secur. **12**(3), 544 (2017)
12. X. Ma, J. Wu, S. Xue, J. Yang, C. Zhou, Q.Z. Sheng, H. Xiong, L. Akoglu, IEEE Trans. Knowl. Data Eng. **35**, 12012–12038 (2021)
13. H. Zhang, L. Cao, P. VanNostrand, S. Madden, E.A. Rundensteiner, in *Proceedings of the ACM SIGKDD, Virtual Event, Singapore, August 14–18, 2021* (2021)
14. D. Han, Z. Wang, W. Chen, Y. Zhong, S. Wang, H. Zhang, J. Yang, X. Shi, X. Yin, in *Proceedings of the CCS 2021, Virtual Event, Republic of Korea, November 15–19, 2021* (2021), pp. 3197–3217

15. C. Fu, Q. Li, K. Xu, in *Proceedings of the NDSS 2023, San Diego, California, USA, February 27–March 3, 2023* (2023)
16. Z. Li, J. Zeng, Y. Chen, Z. Liang, in *Proceedings of the ESORICS 2022, Copenhagen, Denmark, September 26–30, 2022* (2022), pp. 589–609
17. H. Nilforoshan, N. Shah, in *Proceedings of the IEEE DSAA 2019, Washington, DC, USA, October 5–8, 2019* (2019), pp. 351–363
18. D. Rolfsen, *Knots and Links*, vol. 346 (American Mathematical Society, New York, 2003)
19. G. Carlsson, Acta Numer. **23**, 289 (2014)
20. M. Wang, Z. Cang, G. Wei, Nat. Mach. Intell. **2**(2), 116 (2020)
21. R. Milo, S. Shen-Orr, S. Itzkovitz, N. Kashtan, D. Chklovskii, U. Alon, Science **298**(5594), 824 (2002)
22. L. Kovanen, K. Kaski, J. Kertész, J. Saramäki, Proc. Natl. Acad. Sci. **110**(45), 18070 (2013)
23. S. Sun, L. Yu, X. Zhang, M. Xue, R. Zhou, H. Zhu, S. Hao, X. Lin, in *Proceedings of the CCS 2021, Virtual Event, Republic of Korea, November 15–19, 2021* (2021), pp. 287–303
24. C. Jing, C. Wang, C. Yan, in *Proceedings of the Financial Cryptography and Data Security 2019, Frigate Bay, St. Kitts and Nevis, February 18–22, 2019*, vol. 11598 (2019), pp. 588–604
25. A.H. Lashkari, G. Kaur, A. Rahali, in *Proceedings of the ICCNS 2020, Tokyo, Japan, November 27–29, 2020* (2020), pp. 1–13
26. S. Liu, K. Zhang, T. Yao, K. Sheng, S. Ding, Y. Tai, J. Li, Y. Xie, L. Ma, in *Proceedings of the IJCAI 2021, Virtual Event/Montreal, Canada, 19–27 August 2021* (2021), pp. 867–873
27. S. Mahdavifar, N. Maleki, A.H. Lashkari, M. Broda, A.H. Razavi, in *Proceedings of the DASC 2021, Canada, October 25–28, 2021* (2021), pp. 60–67
28. C. Wang, H. Zhu, IEEE Trans. Inf. Forensics Secur. **17**, 2703 (2022)
29. S. Han, X. Hu, H. Huang, M. Jiang, Y. Zhao, in *Proceedings of the NeurIPS 2022*, vol. 35 (2022), pp. 32142–32159
30. S. Hu, X. Zhang, J. Zhou, S. Ji, J. Yuan, Z. Li, Z. Wang, Q. Chen, Q. He, L. Fang, in *Proceedings of the IEEE ICDE 2021, Chania, Greece, April 19–22, 2021* (2021), pp. 2583–2594
31. S.Ö. Arik, T. Pfister, in *Proceedings of the AAAI 2021, Virtual Event, February 2–9, 2021* (2021), pp. 6679–6687
32. G. Pang, A. van den Hengel, C. Shen, L. Cao, in *Proceedings of the ACM SIGKDD 2021, Virtual Event, Singapore, August 14–18, 2021* (2021)
33. A. Abusnaina, M. Abuhamad, H. Alasmary, A. Anwar, R. Jang, S. Salem, D. Nyang, D. Mohaisen, IEEE Trans. Dependable Secure Comput. **19**(5), 3432 (2022)
34. Z. Li, Y. Zhao, J. Han, Y. Su, R. Jiao, X. Wen, D. Pei, in *Proceedings of the ACM SIGKDD 2021, Virtual Event, Singapore, August 14–18, 2021* (2021), pp. 3220–3230
35. Y. Zhu, D. Xi, B. Song, F. Zhuang, S. Chen, X. Gu, Q. He, in *Proceedings of the WWW 2020, Taipei, China, April 20–24, 2020* (2020), pp. 928–938
36. X. Li, X. Li, J. Jia, L. Li, J. Yuan, Y. Gao, S. Yu, IEEE Trans. Inf. Forensics Secur. **18**, 1638 (2023)
37. Y. Ren, H. Zhu, J. Zhang, P. Dai, L. Bo, in *Proceedings of the IEEE ICDE 2021, Chania, Greece, April 19–22, 2021* (2021), pp. 2039–2044
38. Y. Zhang, Y. Fan, Y. Ye, L. Zhao, C. Shi, in *Proceedings of the ACM CIKM 2019, Beijing, China, November 3–7, 2019* (2019), pp. 549–558
39. Y. Liu, X. Ao, Z. Qin, J. Chi, J. Feng, H. Yang, Q. He, in *Proceedings of the WWW 2021, Virtual Event/Ljubljana, Slovenia, April 19–23, 2021* (2021), pp. 3168–3177
40. F. Liu, Y. Wen, D. Zhang, X. Jiang, X. Xing, D. Meng, in *Proceedings of the CCS 2019, London, UK, November 11–15, 2019* (2019), pp. 1777–1794
41. M. Huang, Y. Liu, X. Ao, K. Li, J. Chi, J. Feng, H. Yang, Q. He, in *Proceedings of the WWW 2022, Virtual Event, Lyon, France, April 25–29, 2022* (2022), pp. 1311–1321
42. Z. Liu, C. Chen, X. Yang, J. Zhou, X. Li, L. Song, in *Proceedings of the ACM CIKM 2018, Torino, Italy, October 22–26, 2018* (2018), pp. 2077–2085
43. M.W.A. Nabki, E. Fidalgo, E. Alegre, L. Fernández-Robles, Expert Syst. Appl. **123**, 212 (2019)
44. X. Zhang, Z. Ma, IEEE Trans. Fuzzy Syst. **31**(3), 835 (2023)

45. M. Feng, C. Hsu, C. Li, M. Yeh, S. Lin, in *Proceedings of the WWW 2019, San Francisco, CA, USA, May 13–17, 2019* (2019), pp. 470–479
46. E.P. Xing, A.Y. Ng, M.I. Jordan, S. Russell, in *Proceedings of the NIPS 2002, December 9–14, 2002, Vancouver, British Columbia, Canada* (2002), pp. 505–512
47. E. Welzl, in *New Results and New Trends in Computer Science: Graz, Austria, June 20–21, 1991 Proceedings* (Springer, Berlin, 2005), pp. 359–370
48. A. Singhal, et al., IEEE Data Eng. Bull. **24**(4), 35 (2001)
49. J. Glasser, B. Lindauer, in *2013 IEEE Security and Privacy Workshops* (2013), pp. 98–104
50. L.O. Prokhorenkova, G. Gusev, A. Vorobev, A.V. Dorogush, A. Gulin, in *Proceedings of the NeurIPS 2018, December 3–8, 2018, Montréal, Canada* (2018), pp. 6639–6649
51. G. Pang, C. Shen, A. van den Hengel, in *Proceedings of the ACM SIGKDD 2019, Anchorage, AK, USA, August 4-8, 2019* (2019), pp. 353–362
52. Z. Hu, Y. Dong, K. Wang, Y. Sun, in *Proceedings of the WWW 2020, Taipei, China, April 20–24, 2020* (2020)
53. M.A. Ambusaidi, X. He, P. Nanda, Z. Tan, IEEE Trans. Comput. **65**(10), 2986 (2016)
54. P. Marteau, IEEE Trans. Inf. Forensics Secur. **16**, 2157 (2021)
55. T. Chen, C. Guestrin, in *Proceedings of the ACM SIGKDD 2016, San Francisco, CA, USA, August 13–17, 2016* (2016), pp. 785–794
56. R. Shwartz-Ziv, A. Armon, Inf. Fusion **81**, 84 (2022)
57. P.E. Rauber, A.X. Falcão, A.C. Telea, in *Eurographics Conference on Visualization* (2016), pp. 73–77

Chapter 4
Behavior Molecule-Level Learning on Fine-Grained Attribute Structure

Abstract As machine learning models evolve, simple data no longer meets the information requirements of complex models. Therefore, there is a need for more enriched data to effectively learn intricate patterns. Expanding the underlying information in data presents a nontrivial task as it frequently introduces more noise than useful information. Existing techniques primarily concentrate on treating data as feature vectors, often disregarding the usage of fine-grained global relational information within the data. These global relationships play a pivotal role in augmenting the discriminatory power of data for classification tasks. This chapter introduces a novel method for modeling behavior that incorporates low-noise, fine-grained, structural information. Motivated by the interatomic forces forming chemical molecules, the method captures the global relationships between attributes and models the interactions among internal attributes of behavior to construct a behavior molecule graph for each data point. This approach not only enhances the representational power of the data through structural information but also ensures the mutual isolation of behavior molecule graphs, effectively preventing the propagation of noise. Experimental evaluation on representative datasets from different domains demonstrates the superiority of our fine-grained behavior learning compared to state-of-the-art data classification methods.

4.1 Introduction to Fine-Grained Attribute Structure

Tabular data serves as the medium for recording behavior, providing the foundation for machines to solve real-world problems. Behavior refers to the actions, reactions, or conduct of an entity, whether it is a living organism, a system, or an artificial intelligence agent. Over the past few decades, machine learning-based behavior detection and recognition technologies have witnessed rapid development [1–5]. Detection and recognition of behavior have achieved significant progress across a broad range of applications. By accurately identifying and interpreting behavioral patterns, systems can optimize processes, enhance user experiences, and provide insights for decision support and problem solving in various fields such as finance, healthcare, security, and transportation [6–9].

C. Wang, *Computational Structural Behavior*,
https://doi.org/10.1007/978-981-95-1013-9_4

To identify and utilize behavior, traditional techniques typically represent behavior as feature vectors (e.g., transaction behavior in credit card usage [10], anomalous behavior in network systems [11]), followed by outlier detection using vector spaces [12, 13]. This process involves the extraction, selection, and transformation of features from raw tabular data to ensure that the extracted features better capture the underlying information of the data, thereby enhancing model performance. The dimension of behavior refers to the minimum number of past features required to make optimal predictions about the future within the behavior space [14]. In real-world scenarios, due to constraints such as a low quantity and irrelevant features, observed behavioral features may not fully capture various aspects of behavioral dimensions [15, 16]. Furthermore, with the development of deep learning methods, there is a higher demand for utilizing the rich information present within the data to ensure that the model effectively learns complex patterns underlying the tasks [8, 17].

To improve the informativeness of data and address the requirements of various models, current solutions primarily concentrate on manipulating the data itself, focusing on both the feature perspective and the data volume perspective. Feature extension involves generating new features using specific rules to provide a more comprehensive description of behavior [18–20]. Generative methods use models or algorithms that generate new data samples, thereby augmenting the training dataset and enhancing the performance of machine learning models [21]. Nevertheless, previous methods for enhancing data still have limitations in considering the relationships between data points.

In many fields, graphs are more effective in representing the intricate relationships between data points compared to using feature vectors. This is because graphs offer a direct representation of objects and their connections through nodes and edges [22, 23]. The development of GNNs has brought a new perspective to information enhancement [24]. By modeling the high-order instance feature relations [25], high-order feature interactions [26], and multi-relational correlations among data instances [27], GNNs can appropriately learn graph structures using input data. GNNs learn the potential correlations between data elements, generating robust feature representations that enhance behavior detection and recognition. Due to their ability to capture and model diverse types of information, graphs have found extensive applications across various academic domains, including social activities, biology, e-commerce, academia, and communication [28–32].

While graph-based methods partially leverage the underlying structural information in tabular data, current techniques tend to focus on the relationships between data, constructing structures specifically for node classification tasks [24]. The node representations in these constructed graph structures are primarily updated based on their local neighboring nodes, which limits the modeling of the overall global topological structure and poses challenges in capturing the positional relationships of nodes throughout the entire graph [33, 34]. Furthermore, the construction of graphs for entire datasets inherently involves applying certain rules, which inevitably introduces noisy structures—disordered, interfering, or nonessential patterns within the data, models, or systems. These noisy structures, often caused by external disturbances, measurement errors, incomplete information, or randomness, lack intrinsic meaning and can obscure or distort the recognition of true patterns, behaviors, or relationships. In fields such as machine learning, graph analysis, or sensor networks, such structures can degrade model performance or lead to misjudgments, making denoising, data cleaning, or structure reduction essential. During the learning process of graph structures, the representations of nodes are highly sensitive to the structure, and any noisy structure introduced will be continuously amplified during the message-passing process across the entire graph, consequently affecting the final performance [35].

This chapter aims to address these issues by investigating a novel approach to learning behavioral data, referred to as behavioral molecule-level learning. Our learning approach involves a crucial step of constructing the behavior molecule model (BMM), where a behavior molecule graph is created for each data point based on attribute distribution. The isolated nature of each molecule graph ensures the absence of noise propagation across the data during subsequent learning processes. Furthermore, unlike methods that primarily utilize potential correlations among data, our approach focuses on the more detailed structural information of attributes, enabling further exploitation of the latent information in the data. To learn this fine-grained structural information, we employ graph neural networks for graph-level learning of the data. The primary contributions can be summarized as follows:

- It proposes a novel tabular data representation model, termed the *behavior molecule model*. To our knowledge, this is the first attempt to study the modeling of structural information at the molecular level for behavioral data.
- It designs a novel method for learning behavioral data. The utilization of structural information gain is maximized through graph-level molecular learning.
- It provides the theoretical explanations for the representational capacity of the BMM and the effectiveness of various types of graph neural network aggregations. Furthermore, the experiments conducted on real-world datasets from different domains consistently yielded optimal results, thus validating the effectiveness of the BMM.

4.2 Related Work on Enhancing Information

4.2.1 Behavior Detection and Recognition

The field of behavior analysis is a methodological approach to studying and interpreting individual or system behaviors. With the advancement in network technologies, the study of behavior analysis has expanded from traditional methods such as observation, experimentation, and statistical analysis to more comprehensive intelligent computational analysis utilizing advanced techniques like machine learning [36]. These sophisticated methodologies can be employed to investigate various types of behaviors, ranging from individual behaviors to collective behaviors, and from short-term behaviors to long-term trends [37–39]. Behavior analysis aids in uncovering the causality, patterns, and motivations underlying behaviors, thereby providing robust support for decision-making and problem solving [6–8]. Machine learning methods are utilized for predicting, classifying, and identifying behavior patterns. Traditional machine learning methods rely on feature vectors, which are vectors used to represent data features in machine learning and pattern recognition. With the development of deep learning models, there is a greater demand for information contained within the data to effectively learn complex patterns underlying the data.

4.2.2 Structure-Based Information Enhancement

The most popular supervised learning methods in machine learning conventionally require fixed-size vector inputs, which often results in behavioral data being flattened into tabular formats, overlooking potentially valuable relational information within the data [40]. Recent research has discovered that significant information gains can be achieved by establishing latent graph structures to represent the relationships among behaviors that do not exhibit explicit structures [24].

Generally, data can be structured into three primary forms: homogeneous graphs, heterogeneous graphs, and hypergraphs. In construction methods centered on homogeneous graphs, instance graphs [35, 41], and feature graphs [26, 42] can be constructed using data instances or features as nodes, respectively. For the construction of heterogeneous graphs, [25] connects data instances with their corresponding features (and other metadata) to create bipartite/multipartite graphs. Additionally, [27] treats different feature values as different types of edges, which describe various relationships between data instances, thus constructing a multi-edge/multi-relation graph. Within the framework of hypergraphs, [43] connects table elements with similar attributes via edges, allowing for the linkage of an arbitrary number of table elements within the hypergraph.

After determining the nodes, establishing comprehensive relationships between nodes is crucial for constructing a comprehensive graph structure. Edge construction methods are broadly categorized into four approaches: rule-based methods, learning-based methods, search-based methods, and knowledge-based methods, with the former two being extensively utilized [24]. Rule-based methods either rely on data dependencies between data instances [25, 27, 40] and features or employ common heuristic methods such as k-nearest neighbors [41, 44] or fully connected structures [26, 45]. Learning-based methods treat edges as learnable parameters to generate edges for nodes [46]. Search-based methods utilize neural network search to discover the optimal graph topology for representation learning [47] or employ information retrieval techniques to establish edges for identifying similar relevant data instances [48]. Knowledge-based methods require the involvement of domain experts to supply pertinent knowledge regarding the relationships between data instances, facilitating a comprehensive construction of the graph [49].

4.3 Preliminary

Graph neural networks (GNNs) utilize the inherent graph structure alongside node features X_v to generate a representation vector for each node, denoted as h_v, or for the entire graph, denoted as h_G. Modern GNNs employ a neighborhood aggregation approach, where node representations are updated iteratively by aggregating information from adjacent nodes. Following k iterations of aggregation process, the representation of a node encapsulates the structural nuances present within its k-step network vicinity. Mathematically, the kth layer of a GNN can be formulated as

$$\begin{aligned} a_v^{(k)} &= AGGREGATE^{(k)}\left(\left\{h_u^{(k-1)} : u \in \mathcal{N}(v)\right\}\right), \\ h_v^{(k)} &= COMBINE^{(k)}\left(h_v^{(k-1)}, a_v^{(k)}\right), \end{aligned} \tag{4.1}$$

where the feature vector of node v at the kth iteration/layer is denoted as $h_v^{(k)}$, we initialize $h_v^{(0)}$ as the original feature vector X_v of node v, and $\mathcal{N}(v)$ represents the collection of nodes that are adjacent to node v.

For node classification, the node representation $h_v^{(K)}$ achieved from the final iteration is utilized for prediction purposes. On the other hand, in graph classification, the $READOUT$ function is employed to aggregate the node features obtained from the final iteration, resulting in the overall representation h_G for the entire graph:

$$h_G = READOUT\left(\{h_v^{(K)} \middle| v \in G\}\right). \tag{4.2}$$

The $READOUT$ function can either be a simple permutation invariant function, such as summation, or a more advanced graph-level pooling function.

4.4 Interaction-Based Behavior Molecule Graph

This section elaborates on our proposed behavior fine-grained learning, a novel universal framework for learning behavioral data. We model individual-level structures within the data to explore relationships among attributes. As depicted in Fig. 4.1, behavior molecule-level learning comprises three main components: the attribute discretization module for discretizing general continuous attributes, the molecule graph modeling module for modeling attribute-to-attribute structures, and the graph-level molecule graph learning module for graph-level learning of individual behavior molecule graphs. Within the attribute discretization module, we partition the continuous attribute values, assigning labels to correspond with each attribute value, thereby obtaining a finite number of attribute labels for the dataset. Subsequently, in the molecule graph construct, we use the attribute features of individual data as the feature matrix for the molecule graph. Then, we model

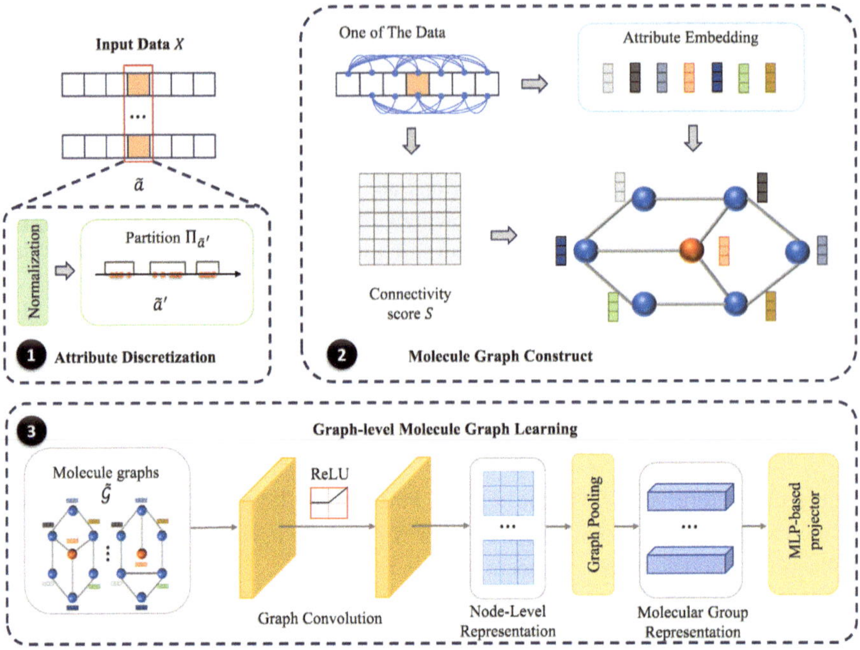

Fig. 4.1 Architecture of the proposed algorithm

the relationships between each attribute of a single data, acquiring an adjacency matrix for each data, resulting in a specific molecule graph corresponding to every data instance. Finally, we employ a graph-level module to perform graph-level task learning on the acquired behavior molecule graphs. The following subsections elucidate these crucial components.

Prior to delving into the specific methods employed, we present a concise summary of the notations used in this part. These notations are provided as preliminary information and are organized in Table 4.1 for easy reference.

4.4.1 *Attribute Discretization*

The nodes in the behavior molecule graph preserve the attribute information of the original data. In real-world scenarios, attributes are typically represented as continuous numerical values within a database. Continuous attributes refer to variables that can assume arbitrary values within a certain range, such as age, height, or temperature. When continuous attributes exist within the data, initializing the node features of the behavior molecule graph directly with the original attributes will lead to an excessive number of node types in the entire molecular group. This may cause the dimensionality of the feature matrix for each behavior molecule

Table 4.1 Notations

Notation	Description
$\widetilde{a}$	A specific continuous attribute set within the behavioral data
$\Pi_{\widetilde{a}}$	A partition on continuous attribute set $\widetilde{a}$
$L_{\widetilde{a}}$	The label set corresponding to each interval in partition $\Pi_{\widetilde{a}}$
X	The behavioral dataset consisting of behavior records
x_k	A behavior record in X
$\mathcal{G} = (A, P)$	A behavior molecule graph consists of a feature matrix P and an adjacency matrix A
$\widetilde{\mathcal{G}}$	The set of behavior molecule graphs
S	The connectivity score matrix
c	The connectivity score bound
$f_\theta(\cdot)$	The GNN-based molecule encoder
H_i	The node representation matrix for $\mathcal{G}_i$
$\widetilde{H}$	The node representation matrices set
$p(\cdot)$	The molecule graph pool converter
U	The molecule group representation matrix
$g_\varphi(\cdot)$	The MLP-based projector
Z	The projected graph representation matrix

graph to explode when conducting graph-level learning. For instance, when height is considered as one of the attributes of behavioral data, an extreme scenario where each data point has a unique height value could occur. If continuous attributes are not discretized, each data point would be assigned a distinct height node, implying that the dimensionality of node features in the molecular graph would at least be equivalent to the number of data points.

Discrete attributes pertain to variables that can only take on distinct values, such as categories or labels. The purpose of attribute discretization is to identify a concise data representation in the form of categories that capture sufficient information from the original continuous attributes, while simplifying the data, reducing problem complexity, and mitigating noise in the data. This approach renders the attributes more interpretable and facilitates their use in learning algorithms, with the ultimate goal of enhancing the generalization capacity of the methods. Prior to modeling the structural aspects of behavioral molecules, it is essential to perform discretization of continuous attributes.

Let $\widetilde{a}$ denote a specific continuous attribute within the behavioral data, with a range of values defined as $[m, t]$. The process of discretizing $\widetilde{a}$ involves identifying a partition of $\widetilde{a}$, denoted as $\Pi_{\widetilde{a}}$, defined as the following set:

$$\Pi_{\widetilde{a}} = \{[m_0, m_1), [m_1, m_2), \ldots, [m_{k-1}, m_k]\}, \tag{4.3}$$

where $m_0 = m$, $m_{i-1} < m_i$, for $i = 1, 2, \ldots, k$, and $m_k = t$.

Each partitioned interval represents a distinct category for the attribute set $\widetilde{a}$. Attributes falling within a specific interval can be represented by the label associated with that interval. Let the set of labels be denoted as $L_{\widetilde{a}}$:

$$L_{\widetilde{a}} = \{l_1, l_2, \ldots, l_k\}. \tag{4.4}$$

The discretization of continuous attributes involves establishing the following mapping, which associates each partitioned interval to its corresponding label: $f : \Pi_{\widetilde{a}} \longmapsto L_{\widetilde{a}}$, for $a \in \widetilde{a}$, if $m_{i-1} \leq a < m_i$, then $f(a) = l_i$.

In this chapter, we employ a clustering method to obtain the partition $\Pi_{\widetilde{a}}$ for the continuous attribute. Subsequently, this partition is utilized to derive labels for each value within the attribute, facilitating subsequent computations. Specifically, we utilize a density-based clustering algorithm, namely the DBSCAN algorithm [50], to perform clustering on continuous attributes. Initially, the continuous attributes underwent data normalization. Let $\widetilde{a} = \{a_1, a_2, \ldots, a_N\}$ denote a continuous attribute. To mitigate the impact of extreme data values, we apply mean normalization as $a_i' = \frac{a_i - \overline{a}}{\sigma}, i = 1, 2, \ldots, N$, where N is the total number of data points, $\overline{a}$ denotes the mean, and σ represents the standard deviation. The resulting normalized attribute set is $\widetilde{a}' = \{a_1', a_2', \ldots, a_N'\}$. Then we set the clustering threshold based on the obtained $\widetilde{a}'$ as

$$\varepsilon = \frac{\max(\widetilde{a}') - \min(\widetilde{a}')}{N}. \tag{4.5}$$

The minimum number of neighbors in the module is denoted as $MinPts$, representing the number of attribute points within a neighborhood that will cluster into a partitioned region. The attribute discretization algorithm is shown in Algorithm 2.

4.4.2 *Molecule Graph Construction*

To extract internal structural information of behaviors, we model them as behavioral molecule graphs. The graph structure is represented as $\mathcal{G} = (\mathcal{V}, \varepsilon, P) = (A, P)$, where $\mathcal{V}$ denotes the set of nodes, ε signifies the set of edges, $P \in \mathbb{R}^{N \times n}$ represents the node feature matrix (with the ith row p_i representing the feature vector of node v_i), and A corresponds to the adjacency matrix. In the behavior molecule graph, nodes represent the attribute information of behaviors, and edges indicate the association relationships between attributes. In the molecule graph modeling module, the calculation of edges relies on the distribution of attribute pairs across the entire dataset. Let X represent the behavioral dataset consisting of behavior records, where each behavior is described by multiple attribute information. $x_k = (x_k^1, x_k^2, \ldots, x_k^i, \ldots, x_k^n)$ is the kth behavior record in X, where x_k^i represents the attribute information at the ith position.

Algorithm 2: The attribute discretization algorithm

Input: A set of attributes $\widetilde{a} = \{a_1, a_2, \ldots, a_N\}$, the distance threshold ε, and the minimum number of points required for a cluster $MinPts$.
Output: A set of discretized attributes $\widetilde{a}' = \{a'_1, a'_2, \ldots, a'_N\}$.

```
for a_i ∈ ã do
    a_i ← (a_i − ā)/a;
end
L ← 0;
for the unvisited a_i ∈ ã do
    mark a_i as visited;
    S ← GETNEIGHBORS(a_i, ε);
    if |S| < MinPts then
        a'_i ← −1;
    end
    else
        a'_i ← L;
        for a_j ∈ S do
            if a_j is not visited then
                mark a_j as visited;
                S' ← GETNEIGHBORS(a_j, ε);
                if |S| ≥ MinPts then
                    S ← S ∪ S';
                end
                a'_j ← L;
            end
        end
    end
    L ← L + 1;
end
return ã';
```

The BMM consists of the feature matrix P and the adjacency matrix A. The feature matrix $P = [p_1, p_2, p_3, \ldots, p_n]$ represents the attribute information of n dimensions in behaviors. The connectivity score matrix $S = (s_{ij})_{n\times n}$ describes the connectivity of the behavior molecule graph. The value of feature matrix P and the connectivity score matrix S can be calculated by the following formula:

$$\begin{cases} p_i = x_k^i, \\ s_{ij} = \dfrac{N_{ij}}{N_i + N_j}, \quad i, j \in \mathcal{N}, \end{cases} \tag{4.6}$$

where $\mathcal{N} = \{1, 2, \ldots, n\}$ and the dimension of behaviors is n, x_k^i represents the attribute of behavior x, N_i represents the number of occurrences of attribute x_k^i of behavior x in the dataset, N_{ij} represents the number of occurrences of both x_k^i and x_k^j in a behavior record in the dataset, and s_{ij} represents the distribution of attribute pairs over the entire dataset, which to some extent represents the association relationship between attributes.

Subsequently, we derive the adjacency matrix $A = (n_{ij})_{n\times n}$ for each behavioral molecule graph based on the connectivity score matrix S. To constrain the size of each molecule graph, we employ a connectivity score bound, denoted as c. Specifically, when $s_{ij} < c$, $n_{ij} = 0$; otherwise, $n_{ij} = 1$.

The calculation for the connectivity score, denoted as s_{ij}, signifies the proportion of co-occurrence for attributes x_k^i and x_k^j relative to their respective individual occurrences. It, to a certain extent, characterizes the relationship between the two attributes. When $s_{ij} < c$, we consider that there is no significant association between the two attributes. Consequently, in the structural graph of behavior x_k, there is no connection between attributes x_k^i and x_k^j. Conversely, when $s_{ij} \geq c$, we infer the presence of an association between them, thus indicating the existence of an edge structure. It is important to note that the presence or absence of an edge represents two discrete states, which can be readily distinguished within the graphical structure.

After applying the aforementioned operations to each behavioral record in X, each record corresponds to a specific structure. Thus, we obtain a set of behavior molecule graphs denoted as $\tilde{\mathcal{G}} = \{\mathcal{G}_1, \mathcal{G}_2, \ldots, \mathcal{G}_N\}$. In each behavior molecule graph, the number of nodes equals the dimensionality of the behavioral record data, i.e., $|\mathcal{V}| = n$. Subsequently, we use a GNN-based module to conduct graph-level learning on the obtained $\tilde{\mathcal{G}}$. The molecule graph construct algorithm is shown in Algorithm 3.

4.4.3 *Graph-Level Molecule Learning*

Given the obtained set of behavior molecule graphs $\tilde{\mathcal{G}}$ corresponding to X, a logical approach is to utilize graph neural networks for graph-level classification tasks. The technical details of graph neural networks are provided in Sect. 4.3. In this module, we use the following components for graph-level behavior molecule graph learning:

GNN-Based Molecule Encoder An augmented graph $\mathcal{G}_i \in \tilde{\mathcal{G}}$ undergoes node-level representation extraction through a GNN-based molecule encoder $f_\theta(\cdot)$:

$$H_i = f_\theta(\mathcal{G}_i), i = 1, 2, \ldots, N, \tag{4.7}$$

where θ represents the parameter of the encoder $f_\theta(\cdot)$, and $H_i \in \mathbb{R}^{n\times d_1}$ (where d_1 denotes the dimension of the representation) represents the node representation matrices for $\mathcal{G}_i$.

Within this component, we employ three distinct aggregation types of $f_\theta(\cdot)$ during the learning process, namely GraphSAGE [51], GCN [52], and GIN [53]. The

Algorithm 3: The molecule graph construct algorithm

Input: A behavioral dataset X. The connectivity score bound c.
Output: A set of behavior molecule graph set $\tilde{\mathcal{G}} = \{\mathcal{G}_0, \mathcal{G}_1, \ldots, \mathcal{G}_N\}$.

```
for x_k ∈ X do
    Initialize an empty graph G_k = (A, P) ∈ G̃;
    for x_k^i ∈ x_k do
        p_i ← x_k^i;
    end
    for x_k^i, x_k^j ∈ x_k do
        if i ≠ j then
            Count the number of times two attributes occur separately and together
              N_i, N_j and N_ij;
            s_ij ← N_ij / (N_i + N_j);
            if s_ij ≥ c then
                a_ij ← 1;
                a_ji ← 1;
            end
        end
    end
    G_k ← (A, P);
end
return G̃;
```

aggregation functions for each component are detailed below. The AGGREGATE function in GraphSAGE is defined as

$$a_v^{(k)} = \mathrm{MAX}\Big(\Big\{\mathrm{ReLU}\Big(W \cdot h_u^{(k-1)}\Big), \forall u \in \mathcal{N}(v)\Big\}\Big). \tag{4.8}$$

Here, MAX represents element-wise max pooling. The COMBINE function in GraphSAGE is

$$h_v^{(k)} = W \cdot \Big[h_v^{(k-1)}, a_v^{(k)}\Big]. \tag{4.9}$$

In GCN, the AGGREGATE and COMBINE functions are integrated as

$$h_v^{(k)} = \mathrm{ReLU}\Big(W \cdot \mathrm{MEAN}\Big\{h_u^{(k-1)}, \forall u \in \mathcal{N}(v) \cup \{v\}\Big\}\Big). \tag{4.10}$$

The AGGREGATE and COMBINE functions of GIN are based on SUM and MLP:

$$h_v^k = \mathbf{MLP}^k\Big((1+\varepsilon^k) \cdot h_v^{k-1} + \sum_{u \in N(v)} h_u^{k-1}\Big). \tag{4.11}$$

Molecule Graph Pool Converter After obtaining the node representation matrices set corresponding to the behavior molecule graph collection, denoted as $\widetilde{H} = \{H_1, H_2, \ldots, H_N\}$, the molecule graph pool converter, denoted as $p(\cdot)$, transforms $\widetilde{H}$ into molecule group representation matrix:

$$U = p(\widetilde{H}) = CONCAT(r(H_1), r(H_2), \ldots, r(H_N)), \tag{4.12}$$

where $U \in \mathbb{R}^{N \times d_1}$ is the molecular group representation matrix for $\widetilde{H}$, $CONCAT$ is a function for vector concatenation, $r(\cdot)$ is a replaceable READOUT function, and here SUM is chosen as the pooling method for the node representation matrices.

MLP-Based Projector After applying the pool converter, the representations are mapped to an alternative latent space using an MLP-based projector $g_\varphi(\cdot)$, where the contrastive loss is computed:

$$Z = g_\varphi(U), \tag{4.13}$$

where φ represents the parameter of the projector $g_\varphi(\cdot)$, and $Z \in \mathbb{R}^{N \times d_2}$ (where d_2 represents the projection dimension) denotes the matrix of projected graph representations for the molecular group representation matrix.

Graph-Level Loss Function For a K-class behavioral dataset, the label set of the data is denoted as $L_X = \{1, 2, \ldots, K\}$. The construction of behavior molecule graphs transforms the labels of the data into the labels of each graph. Each row z_i in the projected graph representation matrix Z, derived from the aforementioned components, encapsulates the structural information of the graph projection. For each projection z_i, we utilize cross-entropy loss as the graph-level loss:

$$\mathscr{L} = -\sum_{k=1}^{K} q(k|z_i) \log p(k|z_i), \tag{4.14}$$

where $p(k|z_i) = \frac{e^{t_k}}{\sum_{j=1}^{K} e^{t_j}}$ represents the probability of z_i calculated by the classifier for label k, and t_j represents the logits, $q(k|z_i)$ is the ground-truth distribution over labels for projection z_j, and $\sum_{k=1}^{K} q(k|z_i) = 1$.

As replaceable component in behavior molecule learning, different graph neural networks exhibit varying performances in learning the BMM. In the following section, we will analyze this module.

4.4.4 Analysis of Different Aggregation Methods

The behavioral data comprises multiple attributes, where the internal structure of behavior constitutes the relationships among these attributes. To analyze the utility brought by this structural information, we simplify the problem to consider the influence of two attributes on the data. Considering the presence of correlated attributes x_1, x_2, y_1, and y_2 within attributes X and Y, let B_1, B_2, B_3, B_4 denote distinct classes of behavioral data falling within these four attributes. Subsequently, focusing on the B_1 class, we analyze the learning process of the BMM. Here, the distances from B_1 to the remaining three classes are denoted as d_{12}, d_{13}, and d_{14}. Clearly, the distance between behaviors within the B_1 class is represented as $d_{11} = 0$.

Considering the perspective of the B_1 class behavior, an explanation is provided for the four classes of behavior:

B_1: Within-class behavioral attributes overlap, indicating a higher likelihood of belonging to the same type of behavior.

B_2/B_3: Only one attribute overlaps with B_1, indicating a relatively lower probability of being the same type of behavior.

B_4: No attribute overlaps with B_1, suggesting a higher probability of being a different type of behavior.

We employ graph neural networks to learn the BMM. Graph neural networks update node features through message-passing mechanisms. For a graph with two nodes, the update of node features is represented as follows:

$$x_1^{(k+1)} = UPDATE\left(x_1^k, y_1^k\right),$$
$$y_1^{(k+1)} = UPDATE\left(x_1^k, y_1^k\right). \tag{4.15}$$

Common graph neural networks such as GIN, GCN, and GraphSAGE can be viewed as employing update mechanisms like sum, average, and max, respectively. Subsequently, we conduct a separate analysis of the message-passing processes within each of the three graph neural networks.

The initial distances between the four classes of behaviors:

$$d_{11}^2 = 0;$$
$$d_{12}^2 = ||y_1 - y_2||_2^2;$$
$$d_{13}^2 = ||x_1 - x_2||_2^2;$$
$$d_{14}^2 = ||x_1 - x_2||_2^2 + ||y_1 - y_2||_2^2. \tag{4.16}$$

SUM Message-Passing After updating the node features, the distances between the four behavior classes are calculated as follows:

$$d_{11}'^2 = 0;$$
$$d_{12}'^2 = 2||y_1 - y_2||_2^2;$$

$$
\begin{aligned}
d_{13}'^2 =& 2||x_1 - x_2||_2^2; \\
d_{14}'^2 =& 2||x_1 - x_2 + y_1 - y_2||_2^2 \\
& < 2\left(||x_1 - x_2||_2^2 + ||y_1 - y_2||_2^2\right).
\end{aligned}
\tag{4.17}
$$

The effect of SUM message-passing is as follows:

$$
\begin{aligned}
& d_{11}'^2 = d_{11}^2;\ d_{12}'^2 = 2d_{12}^2; \\
& d_{13}'^2 = 2d_{13}^2;\ d_{14}'^2 < 2d_{14}^2.
\end{aligned}
\tag{4.18}
$$

It achieves nonuniform expansion, allowing us to observe relative distances. For the B_1 class of intra-class behavioral data, the distance remains unchanged at 0. The distances from B_1 class to B_2 and B_3 classes are amplified to twice the original distance. The distance from B_1 class to B_4 class is enlarged to less than twice the original distance.

Hence, ensuring the differentiation between B_1 class and B_2/B_3 classes, as well as the distinction between B_1 class and B_4 class, is a crucial task. Specifically, the construction method of the structure will significantly impact the identification results.

AVG Message-Passing

$$
\begin{aligned}
d_{11}'^2 =& 0; \\
d_{12}'^2 =& \frac{1}{2}||y_1 - y_2||_2^2; \\
d_{13}'^2 =& \frac{1}{2}||x_1 - x_2||_2^2; \\
d_{14}'^2 =& \frac{1}{2}||x_1 - x_2 + y_1 - y_2||_2^2 \\
& < \frac{1}{2}\left(||x_1 - x_2||_2^2 + ||y_1 - y_2||_2^2\right).
\end{aligned}
\tag{4.19}
$$

The effect of AVG message-passing is as follows:

$$
\begin{aligned}
& d_{11}'^2 = d_{11}^2;\ d_{12}'^2 = \frac{1}{2}d_{12}^2; \\
& d_{13}'^2 = \frac{1}{2}d_{13}^2;\ d_{14}'^2 < \frac{1}{2}d_{14}^2.
\end{aligned}
\tag{4.20}
$$

It achieves nonuniform contraction, allowing us to observe relative distances. For the B_1 class of intra-class behavioral data, the distance remains unchanged at 0. The distances from B_1 class to B_2 and B_3 classes shrink to half of their original distances. The distance from B_1 class to B_4 class reduces to less than half of its original distance.

It is noteworthy that both the SUM and AVG approaches narrow the distances between originally similar nodes while increasing the distances between different nodes. However, given the unchanged intra-class distance of the B_1 class, SUM generally amplifies and increases the differences, whereas AVG generally diminishes and decreases the differences. Theoretically, these two message-passing methods produce distinct final results. In a classification scenario, increasing differences may lead to improved performance.

MAX Message-Passing

$$\begin{aligned} d'^2_{11} &= 0; \\ 0 \leq d'^2_{12} &\leq \frac{1}{2}||y_1 - y_2||_2^2; \\ 0 \leq d'^2_{13} &\leq \frac{1}{2}||x_1 - x_2||_2^2; \\ 0 \leq d'^2_{14} &\leq \frac{1}{2}||x_1 - x_2 + y_1 - y_2||_2^2. \end{aligned} \tag{4.21}$$

GraphSAGE exhibits instability in its states, indicating that different choices of graph neural networks have varying effects on the learning efficiency of the behavioral molecule model.

Overall, GNNs have the capability to decrease distances between originally similar nodes, while increasing distances between different nodes. However, the extent to which nodes of different categories are pushed apart is not consistent. This discrepancy constitutes one of the important reasons leading to divergent final learning outcomes.

4.5 Evaluation of BMM

We analyze the effectiveness of the BMM on multiple real datasets of varying scales and types. Our evaluation focuses on addressing four specific questions:

- **Q1**: In comparison to the state-of-the-art classification methods that do not incorporate structural information and global information, can BMM-GNN achieve superior performance in classification tasks?
- **Q2**: What is the impact of different GNN architectures in BMM-GNN on the final classification performance?
- **Q3**: In BMM, there are two key components, i.e., nodes and edges. What is the contribution of each component to information gain?
- **Q4**: How does the hyperparameter, connectivity score threshold c, affect BMM-GNN?

4.5.1 Experimental Setup

4.5.1.1 Datasets and Evaluation Metric

To demonstrate the effectiveness of our behavioral molecule modeling approach, we conduct the experiments on nine real-world datasets. To ensure generality, these datasets are sourced from various domains, including health and medicine, business, social sciences, among others, obtained from UCI. A detailed description of the datasets is provided in Table 4.2.

To evaluate the performance of the method, we utilize three commonly used metrics: accuracy (Acc), recall (Rec), and F1 score (F1 score).

4.5.1.2 Baseline Methods

We model the data using the BMM and employ three types of graph neural networks for learning: GraphSAGE [51], GCN [52], and GIN [53]. These networks represent the three types of message-passing mechanisms in graph neural networks analyzed in our study, and we compare their performance differences. Simultaneously, we compare them with traditional methods using feature vector representations, including KNN, two types of Bayesian algorithms (GaussianNB, BernoulliNB), two types of decision tree algorithms (DecisionTreeGini, DecisionTreeEntropy), SVM [54], RandomForest [55], XGBoost [56], CatBoost [57], LightGBM [58], and the neural network approach DANet [59]. We also compare our method with LUNAR [44], which leverages global structure methods. By isolating global structural information, we effectively avoid the impact of noise structures. The current state-of-the-art models based on gradient boosting decision tree algorithms and neural network-based methods are shown as follows:

- **XGBoost [56]:** It is an optimized, distributed gradient boosting framework that aims to implement efficient, flexible, and portable machine learning algorithms.

Table 4.2 A summary of datasets, where N, F, and C denote the numbers of samples, features, and classes

Dataset	Domain	N	F	C
Breast cancer	Health and medicine	286	9	2
Credit approval	Business	690	15	2
Glass	Physics and chemistry	214	9	7
Hayes-Roth	Social science	132	5	3
Heart disease	Health and medicine	303	13	2
Ionosphere	Physics and chemistry	351	34	2
Tic-Tac-Toe	Games	958	9	2
Vehicle	Other	846	18	4
Zoo	Biology	101	16	7

- **CatBoost [57]:** It provides adaptive learning rate and categorical feature processing, which control weak learner influence and convert categorical features to numerical ones for better handling and improved model accuracy.
- **LightGBM [58]:** The framework is a high-speed, distributed gradient boosting framework that utilizes decision trees and offers efficient parallel training support.
- **DANet [59]:** It is a neural network approach that provides a tabular data neural component, learning to explicitly group related input features and generate higher-level features for semantic abstraction.
- **LUNAR [44]:** It is a graph neural network-based anomaly detection method that constructs a global graph structure for each dataset. Notably, we utilize GCN and cross-entropy loss to adapt it for multi-classification tasks.

There are three aspects of comparison here. Firstly, the comparison between the BMM with structural information and the original tabular data aims to experimentally verify the impact of the information derived from the data itself in classification tasks. Secondly, compared to methods that use global structures, the BMM achieves molecular-level information processing, effectively addressing the noise associated with using global structures. Finally, different graph neural networks exhibit varying abilities in learning from this structural information. Our theoretical analysis infers distinct utilities of the three classes of message-passing mechanisms in graph neural networks. The experimental section validates our analytical results based on real-world scenarios.

4.5.1.3 Implementation

When processing continuous attribute discretization, we employ a clustering method to obtain the partition Π_A for the continuous attribute. Subsequently, this partition is utilized to derive labels for each value within the attribute, facilitating subsequent computations. Specifically, we utilize a density-based clustering algorithm, namely the DBSCAN algorithm [50], to perform clustering on continuous attributes. Initially, the continuous attributes underwent data normalization. Let $\mathbb{X} = \{x_1, x_2, \ldots, x_n\}$ denote a continuous attribute. To mitigate the impact of extreme data values, we apply mean normalization as follows:

$$x_i' = \frac{x_i - \overline{x}}{\sigma}, i = 1, 2, \ldots, n,$$

where n represents the number of data points, $\overline{x}$ denotes the mean, and σ signifies the standard deviation. The normalized attribute set is $\mathbb{X}' = \{x_1', x_2', \ldots, x_n'\}$. Then we set the clustering threshold based on the obtained $\mathbb{X}'$ as

$$\delta = \frac{\max(\mathbb{X}') - \min(\mathbb{X}')}{n}.$$

The minimum number of nearest neighbors is set to $MinPts = 5$.

After discretizing the continuous attributes of the data, typically resulting in a label count $l_k \ll n$ for each attribute, we proceed with the behavior molecule modeling. When computing the connectivity matrix for the BMM, we choose a connectivity score of $c = 0.1$. In the subsequent ablation experiments, we change the hyperparameter c to explore its impact, representing the correlation between attributes in the data. After modeling the behavior molecule graph for each data entry in the dataset, we employ three separate graph neural networks for graph-level classification of the resulting behavior molecule dataset. To ensure the reliability of our experimental results, the reported values are averaged over 10 experiments.

4.5.2 *Classification Result*

In this section, we compare our classification results in terms of Acc, Rec, and F1 score against the baseline methods (*Q1*). Tables 4.3, 4.4, and 4.5 summarize the average scores of nine datasets. From the table, it is evident that our method achieves the best or second-best results across all datasets. Our comparison is multifaceted. On one hand, compared to traditional methods that utilize feature vectors, our approach models the data as behavior molecule models, thoroughly exploring the latent information within the data. On the other hand, compared to methods that also incorporate structural information, our approach similarly demonstrates advantages, primarily due to its lower noise. Specifically, on the breast cancer, heart disease, ionosphere, and zoo datasets, the best and second-best results are consistently achieved by our method. On the remaining datasets, methods based on feature vectors attain some of the best or second-best results. However, on all datasets, methods based on structural information perform worse than those based on feature vectors, experimentally demonstrating that structural information does not always provide benefits and can introduce certain noise.

Compared to methods that utilize feature vectors, our approach models the data as behavior molecule graphs. This advantage stems from our fine-grained modeling, effectively leveraging latent structural information present in the data. Analogous to the microlevel representation of chemical molecules, our approach signifies the significance of considering structural information from a microscopic perspective in data analysis. Meanwhile, compared to methods that utilize global structural information, our approach offers a significant improvement, which supports our notion that using global information introduces an excessive noise. This noise is amplified through the graph neural network message-passing, as each node connected by a noisy edge is often linked to many other nodes (*Q2*). Additionally, exploring different graph neural networks for learning on the modeled behavior molecule data demonstrates varying learning performance. Overall, these findings align with our analysis section, where we anticipate that "sum" performs better than "mean" and "max" in most cases, and "mean" outperforms the unstable "max" aggregation.

Table 4.3 The comparison of classification accuracy between feature vector-based methods and BMM-based methods on breast cancer, credit approval, and glass

Dataset	Breast cancer			Credit approval			Glass		
Metric	Acc	Rec	F1 score	Acc	Rec	F1 score	Acc	Rec	F1 score
KNN	0.71±0.06	0.31±0.10	0.38±0.10	0.68±0.03	0.58±0.03	0.62±0.04	0.66±0.11	0.60±0.13	0.62±0.13
GaussianNB	0.70±0.04	0.42±0.09	0.44±0.09	0.79±0.02	0.66±0.06	0.74±0.04	0.40±0.07	0.47±0.11	0.51±0.14
BernoulliNB	0.69±0.04	0.22±0.16	0.27±0.16	0.79±0.04	0.73±0.06	0.75±0.05	0.40±0.09	0.31±0.06	0.29±0.11
DecisionTreeGini	0.71±0.05	0.00±0.00	0.00±0.00	0.86±0.03	0.92±0.04	0.85±0.04	0.40±0.12	0.28±0.11	0.20±0.12
DecisionTreeEntropy	0.68±0.05	0.44±0.11	0.44±0.11	0.81±0.04	0.80±0.08	0.79±0.06	0.67±0.06	0.64±0.08	0.68±0.12
SVM	0.69±0.05	0.48±0.11	0.47±0.10	0.69±0.03	0.44±0.06	0.56±0.06	0.59±0.12	0.52±0.14	0.48±0.13
RandomForest	0.74±0.05	0.35±0.11	0.42±0.11	0.87±0.03	0.85±0.06	0.86±0.04	0.76±0.06	0.70±0.07	0.71±0.09
XGBoost	0.75±0.05	0.34±0.10	0.43±0.11	0.87±0.03	0.84±0.06	0.85±0.04	0.73±0.11	0.69±0.11	0.67±0.10
Catboost	0.71±0.05	0.58±0.04	0.57±0.06	0.85±0.02	0.85±0.02	0.85±0.02	0.62±0.07	0.50±0.09	0.50±0.10
LightGBM	0.70±0.05	0.56±0.05	0.56±0.14	0.84±0.03	0.84±0.03	0.84±0.03	0.64±0.09	0.51±0.09	0.54±0.10
DANet	0.73±0.06	0.61±0.07	0.63±0.27	0.85±0.03	0.85±0.04	0.85±0.04	0.67±0.08	0.64±0.09	0.60±0.10
LUNAR	0.72±0.06	0.62±0.18	0.51±0.08	0.83±0.04	0.83±0.04	0.82±0.04	0.45±0.05	0.78±0.09	0.20±0.06
BMM-GraphSAGE	0.76±0.01	0.37±0.03	0.46±0.03	0.88±0.01	0.88±0.01	0.89±0.01	0.53±0.02	0.53±0.02	0.44±0.01
BMM-GCN	0.84±0.01	0.77±0.02	0.78±0.02	0.92±0.02	0.93±0.01	0.91±0.01	0.68±0.02	0.68±0.02	0.62±0.02
BMM-GIN	0.87±0.01	0.82±0.01	0.83±0.02	0.92±0.01	0.93±0.01	0.92±0.01	0.78±0.03	0.78±0.03	0.75±0.03

Table 4.4 The comparison of classification accuracy between feature vector-based methods and BMM-based methods on Hayes-Roth, heart disease, and ionosphere

Dataset	Hayes-Roth			Heart disease			Ionosphere		
Metric	Acc	Rec	F1 score	Acc	Rec	F1 score	Acc	Rec	F1 score
KNN	0.36±0.10	0.38±0.12	0.38±0.11	0.68±0.06	0.47±0.10	0.51±0.08	0.85±0.04	0.81±0.04	0.83±0.04
GaussianNB	0.69±0.09	0.72±0.08	0.69±0.09	0.70±0.13	0.79±0.13	0.66±0.10	0.89±0.04	0.87±0.05	0.88±0.05
BernoulliNB	0.33±0.07	0.33±0.00	0.16±0.03	0.79±0.04	0.65±0.12	0.68±0.08	0.86±0.05	0.84±0.05	0.85±0.06
DecisionTreeGini	0.36±0.08	0.37±0.04	0.23±0.08	0.82±0.05	0.65±0.12	0.72±0.09	0.85±0.05	0.82±0.08	0.82±0.07
DecisionTreeEntropy	0.79±0.09	0.82±0.06	0.81±0.08	0.76±0.06	0.61±0.15	0.63±0.10	0.88±0.03	0.87±0.03	0.87±0.03
SVM	0.57±0.09	0.59±0.10	0.58±0.10	0.77±0.04	0.63±0.10	0.66±0.07	0.92±0.03	0.90±0.04	0.91±0.03
RandomForest	0.77±0.04	0.80±0.04	0.79±0.05	0.83±0.04	0.67±0.11	0.73±0.07	0.92±0.02	0.92±0.02	0.92±0.03
XGBoost	0.76±0.06	0.76±0.07	0.76±0.07	0.80±0.06	0.62±0.13	0.69±0.11	0.92±0.03	0.91±0.03	0.92±0.03
Catboost	0.59±0.14	0.60±0.16	0.60±0.16	0.82±0.05	0.79±0.04	0.79±0.05	0.92±0.02	0.90±0.03	0.91±0.03
LightGBM	0.48±0.06	0.46±0.07	0.46±0.09	0.81±0.05	0.78±0.05	0.79±0.05	0.92±0.03	0.90±0.03	0.91±0.03
DANet	0.78±0.10	0.79±0.10	0.79±0.11	0.79±0.04	0.76±0.05	0.76±0.05	0.90±0.02	0.87±0.03	0.89±0.02
LUNAR	0.53±0.05	0.65±0.12	0.45±0.10	0.78±0.07	0.77±0.07	0.76±0.08	0.67±0.05	0.77±0.11	0.45±0.07
BMM-GraphSAGE	0.80±0.04	0.80±0.04	0.78±0.05	0.85±0.01	0.78±0.03	0.78±0.01	0.89±0.01	0.77±0.03	0.83±0.02
BMM-GCN	0.44±0.04	0.44±0.04	0.31±0.05	0.88±0.01	0.80±0.03	0.83±0.02	0.96±0.01	0.95±0.01	0.95±0.01
BMM-GIN	0.93±0.03	0.94±0.03	0.93±0.03	0.89±0.01	0.84±0.02	0.84±0.02	0.94±0.02	0.93±0.03	0.93±0.03

Table 4.5 The comparison of classification accuracy between feature vector-based methods and BMM-based methods on Tic-Tac-Toe, Vehicle, and Zoo

Dataset	Tic-Tac-Toe			Vehicle			Zoo		
Metric	Acc	Rec	F1 score	Acc	Rec	F1 score	Acc	Rec	F1 score
KNN	0.82±0.03	0.88±0.07	0.86±0.03	0.64±0.03	0.64±0.03	0.63±0.03	0.85±0.08	0.74±0.14	0.70±0.14
GaussianNB	0.69±0.01	0.90±0.04	0.79±0.01	0.46±0.03	0.46±0.02	0.43±0.03	0.92±0.06	0.92±0.06	0.83±0.14
BernoulliNB	0.67±0.05	0.91±0.07	0.78±0.02	0.22±0.03	0.25±0.01	0.22±0.12	0.89±0.07	0.83±0.11	0.78±0.13
DecisionTreeGini	0.64±0.03	1.00±0.00	0.78±0.02	0.23±0.03	0.25±0.00	0.09±0.01	0.70±0.12	0.52±0.10	0.44±0.08
DecisionTreeEntropy	0.92±0.02	0.93±0.02	0.93±0.02	0.75±0.04	0.75±0.04	0.75±0.04	0.88±0.07	0.82±0.10	0.78±0.11
SVM	0.96±0.02	0.97±0.02	0.97±0.01	0.82±0.03	0.82±0.03	0.82±0.02	0.94±0.04	0.88±0.10	0.86±0.11
RandomForest	0.95±0.02	0.99±0.02	0.96±0.02	0.78±0.03	0.79±0.02	0.78±0.02	0.93±0.05	0.85±0.14	0.83±0.15
XGBoost	0.97±0.02	0.98±0.02	0.98±0.01	0.78±0.02	0.78±0.02	0.78±0.02	0.94±0.07	0.89±0.12	0.88±0.13
Catboost	0.84±0.04	0.78±0.05	0.81±0.05	0.68±0.03	0.70±0.02	0.67±0.02	0.88±0.10	0.82±0.13	0.75±0.15
LightGBM	0.84±0.05	0.77±0.07	0.79±0.07	0.73±0.03	0.75±0.03	0.73±0.03	0.73±0.09	0.58±0.12	0.51±0.15
DANet	0.97±0.02	0.97±0.02	0.97±0.02	0.83±0.03	0.84±0.03	0.84±0.03	0.88±0.07	0.84±0.07	0.75±0.13
LUNAR	0.67±0.03	0.65±0.11	0.50±0.07	0.42±0.05	0.49±0.08	0.35±0.07	0.65±0.07	0.68±0.13	0.32±0.07
BMM-GraphSAGE	0.86±0.03	0.71±0.09	0.75±0.07	0.56±0.02	0.56±0.02	0.51±0.02	0.72±0.04	0.72±0.04	0.63±0.04
BMM-GCN	0.88±0.03	0.72±0.09	0.79±0.06	0.65±0.01	0.65±0.01	0.64±0.01	0.98±0.01	0.98±0.01	0.97±0.02
BMM-GIN	0.99±0.00	0.98±0.01	0.98±0.01	0.84±0.02	0.84±0.02	0.84±0.02	0.96±0.02	0.96±0.02	0.96±0.02

Considering the presence of homophilic and heterophilic structures within graph datasets, these two concepts are critically important in understanding the graph structure. Homophily refers to the tendency of nodes with similar attributes or features to connect or form links with each other, thereby creating clusters of similar nodes. Conversely, heterophily indicates the trend of nodes with dissimilar attributes or features to connect within the graph, leading to connections between diverse nodes.

We speculate that certain datasets might exhibit underlying heterophilic properties, which can significantly influence the performance of various aggregation methods used in graph neural networks. For instance, the ionosphere and zoo datasets might demonstrate such heterophilic characteristics. This underlying heterophily could potentially enhance the performance of aggregation methods such as "mean" and "max" aggregations, which are generally considered less stable in homophilic settings.

For further understanding, we present a sample visualization of the behavior molecule graphs corresponding to data points in the breast cancer dataset (see Fig. 4.2). The number of nodes in each molecular graph matches the number of attributes describing behavior in the dataset, with edges representing the interactions between attributes. Through this modeling process, complex and diverse fine-grained structural information is integrated with the behavioral data. Simultaneously, the isolation of behavior molecule models prevents the spread of structural noise, achieving low-noise fine-grained information gain.

In the future work, we aim to delve deeper into the investigation of these potential properties within different datasets. Understanding the extent and impact of heterophily could provide valuable insights for designing more robust and adaptive graph neural network models. This exploration might also lead to the development of new aggregation strategies that are better suited to handle the diversity in node connections, thereby improving the overall performance and stability of models across various datasets.

4.5.3 Ablation Study

In the BMM, two pivotal components encapsulate information (*Q3*). On the one hand, post-discretization, the nodes retain attribute information from the original data; on the other hand, the structure embodies potential microstructural information within the data. Therefore, we proceed to conduct ablation studies on these two components to individually verify their roles within the BMM.

Edge We conduct an ablation study on the structural information to assess the information gain derived from microstructural details present in the data. Specifically, when the edges are removed, the BMM reduces to a collection of discrete points. Each behavior molecule graph is composed of the original feature matrix along with a unit matrix, where there is no message-passing between nodes (original

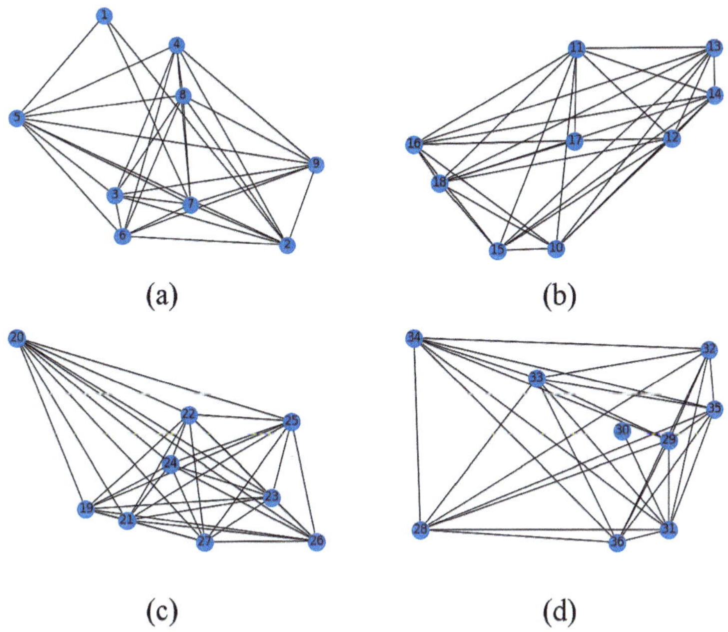

Fig. 4.2 Visualization of behavior molecule graphs corresponding to data points in the breast cancer dataset

attributes). The results of the ablation experiments are depicted in Table 4.6. Across all experiments, the removal of structural information consistently leads to a decline in performance. This indicates that the microstructural information obtained from the original data acts as an information gain, enhancing the model's performance. It highlights the essentiality of structural information.

Node We conduct an ablation study on node information to verify the necessity of preserving original information within the graph structure. Discretization of continuous attributes is necessary in this step. If continuous attributes are included in the nodes, the initialization of node embeddings during graph learning would result in an excessive number of node categories, leading to a dimension explosion. This limitation would render the machine unable to learn within the constraints of existing computational resources. Hence, we discretize continuous attributes into a limited number of labels, seemingly causing some information loss. However, this approach significantly enhances the generalization capability of the modeling method. In this segment, we erase the original node information from the BMM and utilize the degree matrix as the initial embedding of nodes. The results are compared with the modeling method that retains the original attribute information, as illustrated in Table 4.6.

It is noteworthy that after removing the original attribute information, the model's performance experiences a significant decline. In most cases, the performance

Table 4.6 Experiments on node and edge ablation in the BMM model aim to validate the amount of information carried by each of these components

Metric	Acc	Rec	F1 score	Acc	Rec	F1 score	Acc	Rec	F1 score
Dataset	Breast cancer			Credit approval			Glass		
BMM-Full	**0.8720**	**0.8194**	**0.8313**	**0.9174**	**0.9260**	**0.9250**	**0.7763**	**0.7763**	**0.7499**
Edge ablation	0.7562	0.6489	0.6465	0.8444	0.8428	0.8247	0.6943	0.4634	0.4208
Node ablation	0.7206	0.6231	0.6263	0.6483	0.6438	0.6406	0.5489	0.4165	0.3933
Dataset	Hayes-Roth			Heart disease			Ionosphere		
BMM-Full	**0.9282**	**0.9368**	**0.9284**	**0.8903**	**0.8419**	**0.8427**	**0.9416**	**0.9303**	**0.9290**
Edge ablation	0.9110	0.9267	0.9102	0.8384	0.8153	0.8059	0.7862	0.7310	0.6898
Node ablation	0.4991	0.5915	0.4437	0.7483	0.7081	0.7092	0.8762	0.8585	0.8586
Dataset	Tic-Tac-Toe			Vehicle			Zoo		
BMM-Full	**0.9890**	**0.9790**	**0.9838**	**0.8365**	**0.8365**	**0.8362**	**0.9608**	**0.9608**	**0.9588**
Edge ablation	0.9377	0.9281	0.9282	0.6710	0.6773	0.6419	0.8041	0.7069	0.6185
Node ablation	0.9282	0.9102	0.9173	0.4732	0.4739	0.4669	0.7946	0.7318	0.6305

decrease surpasses that observed in the edge ablation. This observation indicates that in the training and learning process of the BMM, the primary influence comes from the original attribute information of the behaviors. The structural information plays a supplementary role in enhancing information gain.

4.5.4 Parameter Analysis

In our behavior molecule modeling, the connectivity score threshold c governs the number of edges in each behavior molecule graph, making it a crucial hyperparameter in the modeling process (*Q4*). When $c = 0$, the molecule model becomes a fully connected graph, and the classification solely relies on the influence of attribute information propagation. To validate its impact in experiments, we adjust the value of c, and the results are shown in Fig. 4.3. It can be observed that with the increase in the value of c, all time expenditures decrease. This is because the connectivity score boundary restricts the size of each behavior molecule graph, significantly reducing the computational costs. Simultaneously, this process retains the strong correlations between attributes in the behavior molecule graph, typically with fewer attributes exhibiting strong correlations. Hence, when the value of c becomes excessively large, the performance tends to decline. Moreover, during experimentation, we observe that selecting a large c value results in the disappearance of the graph structure. At this point, each behavior molecule graph retains only discrete attribute nodes, causing the behavioral molecular model to degrade into a typical form of unstructured feature vectors.

Therefore, our selection of the connectivity score boundary c results in the following observations. On one hand, to ensure the experimental performance and retain as many interrelated attributes' structures as possible, the value of c

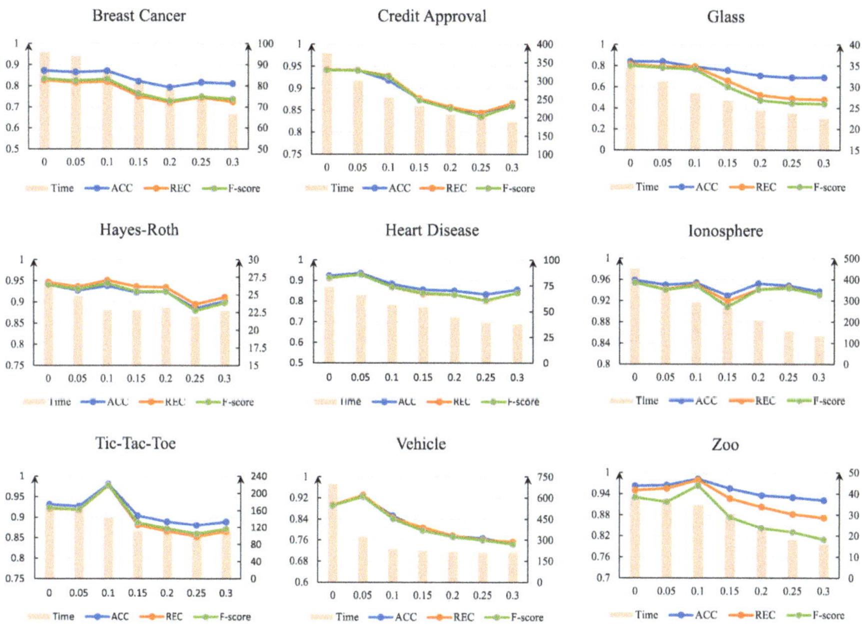

Fig. 4.3 The impact of connectivity score (c) on performance

should not be excessively large. On the other hand, preserving an excessive amount of structure in the behavior molecule graph, although aiming to retain complete structural information, often leads to a significant computational burden. Hence, under conditions of sufficient computational resources, a smaller value for the hyperparameter c would strike a balance between performance and computational expenses.

4.6 Conclusion

We aim to explore the internal structural information of the tabular data to enhance information gain for data classification. We construct a behavior molecule model, utilizing attribute distributions in the data to model relationships among attributes. To ensure generalization, continuous attributes are discretized for representation. For our BMM, we conduct an analysis of performance across various types of graph neural networks. The experiments on real-world datasets from diverse domains demonstrate the effectiveness of mining intrinsic structural information within the data. Due to the outstanding performance of fine-grained structural information gain in classification tasks, our future work will focus on leveraging generative models to further model the intrinsic structure of behavioral data.

References

1. R. Bost, R.A. Popa, S. Tu, S. Goldwasser, in *Proceedings of the NDSS, San Diego, California, USA, August 2015* (2015)
2. Y. Wang, J. Hu, in *Proceedings of the WWW, Honolulu, Hawaii, USA, May 2002* (2002) pp. 242–250
3. I. Firdausi, A. Erwin, A.S. Nugroho, et al., in *Proceedings of the IEEE ACT, 2010* (2010), pp. 201–203
4. T. Shon, J. Moon, Inf. Sci. **177**(18), 3799 (2007)
5. X. Ma, J. Wu, S. Xue, J. Yang, C. Zhou, Q.Z. Sheng, H. Xiong, L. Akoglu, IEEE Trans. Knowl. Data Eng. **35**(12), 12012 (2021)
6. S. Mahmoud, A. Lotfi, C. Langensiepen, Appl. Soft Comput. **13**(4), 1813 (2013)
7. W. Liu, W. Luo, D. Lian, S. Gao, in *Proceedings of the IEEE CVPR 2018, Salt Lake City, UT, USA, June 2018* (2018), pp. 6536–6545
8. M. Reichstein, G. Camps-Valls, B. Stevens, M. Jung, J. Denzler, N. Carvalhais, F. Prabhat, Nature **566**(7743), 195 (2019)
9. T. Li, Z. Hong, Q. Cai, L. Yu, Z. Wen, R. Yang, IEEE Trans. Knowl. Data Eng. **35**(12), 12109 (2021)
10. S. Xuan, G. Liu, Z. Li, L. Zheng, S. Wang, C. Jiang, in *Proceedings of the IEEE ICNSC, 2018* (2018), pp. 1–6
11. P. Horchulhack, E.K. Viegas, A.O. Santin, Comput. Netw. **202**, 108618 (2022)
12. G. Pang, C. Shen, L. Cao, A.V.D. Hengel, ACM Comput. Surv. **54**(2), 1 (2021)
13. N. Elmrabit, F. Zhou, F. Li, H. Zhou, in *Proceedings of the IEEE Cyber Security, 2020* (2020), pp. 1–8
14. W. Bialek, Proc. Nat. Acad. Sci. **119**(18), e2021860119 (2022)
15. Y. Dang, Q. Lin, P. Huang, in *Proceedings of the IEEE/ACM ICSE, Montreal, QC, Canada, May 2019* (2019), pp. 4–5
16. C.M. Booth, S. Karim, W.J. Mackillop, Nat Rev Clin Oncol. **16**(5), 312 (2019)
17. Q. Zhang, L.T. Yang, Z. Chen, P. Li, Inf. Fusion **42**, 146 (2018)
18. L. Sang, F. Xie, X. Liu, X. Wu, in *Proceedings of the IEEE ICDMW, Barcelona, Spain, December 2016* (2016), pp. 677–683
19. Y. Xu, Z. Yu, W. Cao, C.P. Chen, IEEE Trans. Knowl. Data Eng. **35**(1), 16 (2021)
20. W. Zhuge, F. Nie, C. Hou, D. Yi, IEEE Trans. Knowl. Data Eng. **29**(10), 2347 (2017)
21. A. Oussidi, A. Elhassouny, in *Proceedings of the IEEE ISCV, 2018* (2018), pp. 1–8
22. P. Veličković, Curr. Opin. Struct. Biol **79**, 102538 (2023)
23. Z. Li, M. Jiang, S. Wang, S. Zhang, Drug Discov. Today, **27**(12), 103373 (2022)
24. C.T. Li, Y.C. Tsai, J.C. Liao, in *Proceedings of the IEEE ICDE, Anaheim, CA, USA, April 2023* (2023), pp. 3589–3592
25. J. You, X. Ma, Y. Ding, M.J. Kochenderfer, J. Leskovec, Proc. Adv. Neural Inf. Process. Syst. **33**, 19075 (2020)
26. Z. Li, Z. Cui, S. Wu, X. Zhang, L. Wang, in *Proceedings of the ACM CIKM, Beijing, China, November 2019* (2019), pp. 539–548
27. X. Guo, Y. Quan, H. Zhao, Q. Yao, Y. Li, W. Tu, arXiv preprint arXiv:2108.09127 (2021)
28. Z. Liu, J.X. Yu, Y. Ke, X. Lin, L. Chen, in *Proceedings of the IEEE ICDM, Pisa, Italy, December 2008* (2008), pp. 917–922
29. J. Wu, Z. Hong, S. Pan, X. Zhu, C. Zhang, Z. Cai, in *Proceedings of the SIAM, Philadelphia, Pennsylvania, USA, April 2014* (2014), pp. 217–225
30. L. Gao, J. Wu, C. Zhou, Y. Hu, in *Proceedings of the AAAI, San Francisco, California, USA, February 2017*, vol. 31 (2017)
31. C.C. Aggarwal, Y. Zhao, S.Y. Philip, in *Proceedings of the IEEE ICDE, Hannover, Germany, April 2011* (2011), pp. 399–409
32. J. Wu, S. Pan, X. Zhu, C. Zhang, S.Y. Philip, IEEE Trans. Neural Networks Learn. Syst. **29**(7), 3236 (2017)

33. M. Xu, H. Wang, B. Ni, H. Guo, J. Tang, in *Proceedings of the ICML, Virtual Event, July 2021* (2021), pp. 11,548–11,558
34. W.X. Sun, H. Xue, Knowl-based Syst. **230**, 107383 (2021)
35. Y. Liu, Y. Zheng, D. Zhang, H. Chen, H. Peng, S. Pan, in *Proceedings of the WWW, Virtual Event, Lyon, France, April 2022* (2022), pp. 1392–1403
36. M.I. Jordan, T.M. Mitchell, Science **349**(6245), 255 (2015)
37. L. Tang, X. Wang, H. Liu, IEEE Trans. Knowl. Data Eng. **24**(6), 1080 (2012)
38. C. Gao, X. He, D. Gan, X. Chen, F. Feng, Y. Li, T.S. Chua, L. Yao, Y. Song, D. Jin, IEEE Trans. Knowl. Data Eng. **33**(6), 2588 (2019)
39. S. Yang, D. Samarasinghe, A. Arukgoda, S. Abpeikar, E. Lakshika, M. Barlow, IEEE Access **11**, 89077 (2023)
40. M. Cvitkovic, arXiv preprint arXiv:2002.02046 (2020)
41. E. Rocheteau, C. Tong, P. Veličković, N. Lane, P. Liò, arXiv preprint arXiv:2101.03940 (2021)
42. M. Kim, H.S. Choi, J. Kim, arXiv preprint arXiv:2204.03225 (2022)
43. D. Cai, C. Sun, M. Song, B. Zhang, S. Hong, H. Li, in *Proceedings of the SIAM SDM, Alexandria, VA, USA, April 2022* (2022), pp. 127–135
44. A. Goodge, B. Hooi, S.K. Ng, W.S. Ng, in *Proceedings of the AAAI, March 2022* (2022), vol. 36, pp. 6737–6745
45. I. Spinelli, S. Scardapane, A. Uncini, Neural Netw. **129**, 249 (2020)
46. Y. Dou, Z. Liu, L. Sun, Y. Deng, H. Peng, P.S. Yu, in *Proceedings of the CIKM, Ireland, October 2020* (2020), pp. 315–324
47. Y. Xie, Z. Wang, Y. Li, B. Ding, N.M. Gürel, C. Zhang, M. Huang, W. Lin, J. Zhou, in *Proceedings of the ACM SIGKDD, Singapore, August 2021* (2021), pp. 3795–3805
48. K. Du, W. Zhang, R. Zhou, Y. Wang, X. Zhao, J. Jin, Q. Gan, Z. Zhang, D.P. Wipf, Proc. Adv. Neural Inf. Process. Syst. **35**, 16373 (2022)
49. L. Du, F. Gao, X. Chen, R. Jia, J. Wang, J. Zhang, S. Han, D. Zhang, in *Proceedings of the ACM SIGKDD, Singapore, August 2021* (2021), pp. 322–331
50. M. Ester, H.P. Kriegel, J. Sander, X. Xu, et al., in *Proceedings of the KDD-96, Portland, Oregon, USA*, vol. 96 (1996), pp. 226–231
51. W. Hamilton, Z. Ying, J. Leskovec, Proc. Adv. Neural Inf. Process. Syst. **30**, 1024–1034 (2017)
52. T.N. Kipf, M. Welling, in *Proc. ICLR, Toulon, France, April 2017* (2017)
53. K. Xu, W. Hu, J. Leskovec, S. Jegelka, in *Proc. ICLR, New Orleans, LA, USA, May 2019* (2019)
54. C. Cortes, V. Vapnik, Mach. Learn. **20**, 273 (1995)
55. L. Breiman, Mach. Learn. **45**, 5 (2001)
56. T. Chen, C. Guestrin, in *Proceedings of the ACM SIGKDD, San Francisco, CA, USA, August 2016* (2016), pp. 785–794
57. A.V. Dorogush, V. Ershov, A. Gulin, arXiv preprint arXiv:1810.11363 (2018)
58. G. Ke, Q. Meng, T. Finley, T. Wang, W. Chen, W. Ma, Q. Ye, T.Y. Liu, in *Advances in Neural Information Processing Systems*, vol. 30, pp. 3146–3154 (2017)
59. J. Chen, K. Liao, Y. Wan, D.Z. Chen, J. Wu, in *Proceedings of the AAAI, March 2022*, vol. 36 (2022), pp. 3930–3938

Chapter 5
Improving Behavioral Data by Behavioral Molecule Generation

Abstract High-quality real-world data is a critical element for training accurate machine learning models. Challenges such as privacy concerns and information constraints often result in incomplete or insufficient datasets, limiting the ability of models to fully capture underlying behaviors. Generative models have demonstrated great potential in synthesizing high-quality data, particularly for tabular datasets. Nevertheless, due to the complexity of subtle relationships between attributes in tabular data, generating high-quality synthetic tabular data remains a significant challenge. Inspired by computer-aided drug design, this chapter proposes a novel generative model, TabBMG, which integrates inter-attribute relationships within tabular data. It also introduces a unified attribute binning method to handle mixed data types. Experimental results demonstrate that data generated by TabBMG, when combined with graph neural networks, achieves superior classification performance compared to state-of-the-art methods.

5.1 Introduction to Behavioral Molecule Generation

Tabular data records have played a crucial role as carriers of information in machine learning, attracting widespread attention [1, 2]. High-quality real-world data constitutes a critical component for training precise models [3, 4]. Behavioral science employs the concept of dimensions to describe the minimum set of features necessary to characterize a specific behavior, while also serving as a framework to describe the structure of behavior [5]. A single attribute in tabular data can be considered a description of one dimension of behavior. However, due to issues such as privacy protection and information constraints, observable behavioral features may not fully represent all dimensions of behavior [6–8]. In other words, the behavioral features that we can perceive and measure may be incomplete or insufficient to capture all aspects of the underlying behavior. Generative models, which can synthesize high-quality data and learn the underlying structures of complex datasets, have attracted substantial attention in the field of deep learning. These models have been successfully applied to various data types, including video [9], images [10], text [11], and tabular data [12]. Consequently, as generative methods continue to be

C. Wang, *Computational Structural Behavior*,
https://doi.org/10.1007/978-981-95-1013-9_5

extensively studied, leveraging deep generative approaches for synthesizing tabular data emerges as an effective solution to address these challenges [13–15].

The primary objective of synthetic data generation is to produce training datasets that closely resemble real-world data, enabling models trained on synthetic data to perform effectively on real-world tasks [15]. However, recent studies have revealed that low-quality synthetic data can adversely affect model performance, leading to decreased accuracy [16]. Generating high-quality synthetic data that captures the nuanced intricacies of real-world datasets remains a significant challenge [17, 18]. Despite notable advancements in generative modeling for synthetic data, these models often struggle to replicate the complex subtleties of real-world data, particularly when handling noisy, mislabeled, or biased datasets [15]. Moreover, the inherent heterogeneity of individual features and the relatively small size of typical tabular datasets pose additional challenges for training high-quality models for tabular data, making this task more complex compared to fields like computer vision or natural language processing [14].

To address the challenge of high-quality tabular data synthesis, we draw inspiration from computer-aided drug design. In this domain, effective molecular representation is a critical task [19, 20]. Sequence-based representations, such as the simplified molecular input line entry system (SMILES) [21], offer a convenient and straightforward format for encoding molecular structures. However, SMILES representations fall short in capturing the complex spatial relationships among atoms within molecules, rendering them insufficient for predicting certain molecular properties [22, 23]. In contrast, the structural information embedded in molecular graph data provides significant advantages for predicting molecular properties, as it allows for a more effective representation and analysis of information related to drug molecules [24, 25]. Similarly, tabular data, as one-dimensional data, often exhibits relationships among its attributes. Unlike the relationships among individual data entries (such as euclidean distances or similarities), the relationships among attributes within a single data record resemble the interatomic connections within a molecule. These attribute relationships constitute the internal structure of tabular data, much like interatomic bonds define the internal structure of a molecule. With the recent advancements in graph neural networks (GNNs), it has become increasingly feasible to directly and effectively aggregate information from neighboring nodes. This capability is particularly beneficial for capturing the spatial relationships between nodes [26, 27], making GNNs a promising tool for modeling the intricate relationships within tabular data.

Unlike the limited types of atoms that constitute molecules, tabular data typically comprises mixed data types, with attributes consisting of both continuous and discrete variables [14, 28]. Various methods have been proposed to handle continuous variables, such as normalizing continuous values using variational Gaussian mixture models [13] or applying logarithmic transformations to address long-tailed distributions [29]. For discrete variables, the most common approach involves sampling in continuous space after one-hot encoding, sometimes using techniques like Gumbel-Softmax [12, 13]. To effectively manage these two distinct types of attributes, we adopt an information aggregation strategy. Discrete variables are unified with continuous variables through a process that aggregates them into representative attribute clusters, effectively implementing an attribute binning mechanism. This approach facilitates a more coherent representation of tabular data, while accommodating the unique characteristics of its mixed data types.

The main contributions of this work are as follows:

- We introduce a novel generative model for tabular data, termed TabBMG, which leverages behavior molecule generation to model and synthesize tabular datasets. To the best of our knowledge, this is the first approach that integrates inter-attribute relationships into the process of tabular data generation.
- To address the challenges posed by mixed attribute types, we design a unified attribute binning method that effectively aggregates information across heterogeneous attributes.
- Using data generated by TabBMG, classification performance with graph neural networks surpasses state-of-the-art methods, demonstrating the efficacy of the proposed approach.

5.2 Related Work on Generating Data

Generative models for tabular data have become an active area of research in the field of machine learning, as many tabular tasks require high-quality synthetic data. Unlike in visual or natural language processing (NLP) tasks, where vast amounts of extra data are available from the Internet, tabular datasets are typically limited in size. Early research on synthetic data generation modeled each column of a table as a random variable and learned the joint multivariate probability distribution, from which samples were drawn to generate new data. For example, a set of discrete variables might be modeled using decision trees [30] or Bayesian networks [31, 32], while a set of nonlinearly correlated continuous variables could be modeled using copulas [33, 34]. These models, however, were constrained by distribution types and computational issues, severely limiting the fidelity of the generated data.

With the continuous development of generative modeling techniques, researchers have turned to methods like variational autoencoders (VAE) and generative adversarial networks (GANs) to generate tabular data [13, 35, 36]. CTGAN and TVAE [13] introduced preprocessing methods for handling column types in tabular data, employing Gumbel-Softmax to approximate discrete variables in continuous space. OCT-GAN [36], based on neural ODEs, is another generative model aimed at tabular data. More recently, with the rise of diffusion models, their potential for data generation has been increasingly recognized, and numerous studies have applied them to tabular data generation, achieving state-of-the-art results [14, 28]. For instance, CoDi [14] uses two diffusion models to separately process continuous and discrete variables in tabular data, with the models evolving together through mutual conditioning. TabDDPM [28] is a more general approach applicable to any tabular dataset, capable of handling various feature types. However, the aforementioned methods treat tabular data as discrete feature vectors, overlooking the potential relationships between attributes. Recent research has shown that these inter-attribute relationships can serve as a valuable source of information gain for classification tasks involving tabular data [37, 38]. Therefore, we first employ an aggregation method to bin attributes, thereby achieving the aggregation of information and unifying continuous and discrete attributes. We then utilize a behavior molecule generation model to model tabular data, effectively capturing the inter-attribute relationships embedded within the data.

5.3 Behavior Molecule Generation Model

This section provides a detailed description of the proposed behavior molecule-based tabular data generation model, which incorporates information aggregation and dimensionality enhancement through modality transformation to achieve high-quality generation of real-world data.

As depicted in Fig. 5.1, our model consists of three components: the behavioral attribute binning module for aggregating information, the tabular data generation module for producing latent data, and the behavior molecule construction module for implementing modality transformation and dimensionality enhancement. In the behavioral attribute binning module, we first perform attribute aggregation and binning on the raw data, unifying attributes of different types to facilitate information aggregation, thereby obtaining high-quality, unified attribute types. Subsequently, in the tabular data generation module, we use a generative adversarial model to simulate latent data, generating the initial modality of tabular data. Finally, through the behavior molecule construction module, we implement modality transformation and dimensionality enhancement to produce high-quality synthetic data, which is then used for graph-level task learning. The following subsections provide a detailed explanation of these key components.

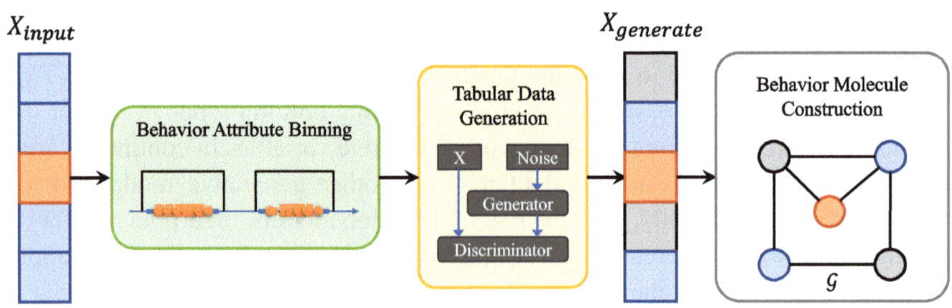

Fig. 5.1 Architecture of the proposed TabBMG

5.3.1 Behavioral Attribute Discretization

In the real world, certain attributes of behavior are often represented as continuous values within a specific range. Continuous attributes are variables that can take any value within a specific range, such as age, height, or temperature. When continuous attributes are present in the data, the distribution of attributes and behavior types often shows correlations within a defined region. However, in some cases, these correlations can be ambiguous and susceptible to the influence of boundary values. To address this, we employ an attribute binning method for initial aggregation and denoising of the attributes.

Moreover, when performing behavior molecule modeling, directly initializing the node features of the behavior molecule graph with raw attributes can lead to an excessive number of node types within the entire molecular group. During graph-level learning, this could result in an explosion in the dimensionality of the feature matrix for each behavior molecule graph. For instance, when weight is considered as one of the attributes of behavioral data, extreme cases may arise where each data point has a unique value. Without attribute binning, each data point would be assigned a distinct node, meaning the dimensionality of the node features in the molecular graph would be at least equal to the number of data points.

Discrete attributes, on the other hand, take values from a finite set, such as categories or labels. The purpose of behavioral attribute binning is to identify a concise data representation in the form of categories, capturing sufficient information from the original continuous attributes while simplifying the data, reducing the complexity of the problem, and mitigating noise in the data. This approach enhances the interpretability of the attributes and facilitates their use in learning algorithms, with the ultimate goal of improving the generalization capability of the method.

Let $\widetilde{a}$ represent a specific attribute in the tabular data, with its value range defined as $[m, t]$. Attribute binning involves partitioning this range into intervals, forming a set $\Pi_{\widetilde{a}}$, defined as:

$$\Pi_{\widetilde{a}} = \{[m_0, m_1), [m_1, m_2), \ldots, [m_{k-1}, m_k]\}, \tag{5.1}$$

where $m_0 = m$, $m_{i-1} < m_i$, for $i = 1, 2, \ldots, k$, and $m_k = t$.

Each interval in $\Pi_{\widetilde{a}}$ is mapped to a corresponding label using the function $f : \Pi_{\widetilde{a}} \longmapsto L_{\widetilde{a}}$. Specifically, for $a \in \widetilde{a}$, if $m_{i-1} \leq a < m_i$, then $f(a) = l_i$.

To construct the partition $\Pi_{\widetilde{a}}$ for continuous attributes, we employ a density-based clustering algorithm, DBSCAN [39]. The clustering process begins with data normalization to mitigate the impact of extreme values. Let $\widetilde{a} = a_1, a_2, \ldots, a_N$ represent a continuous attribute with N data points. Normalization is performed using mean normalization, defined as: $a_i' = \frac{a_i - \overline{a}}{\sigma}, i = 1, 2, \ldots, N$, where $\overline{a}$ is the mean, and σ is the standard deviation. The normalized attribute set is denoted as $\widetilde{a}' = a_1', a_2', \ldots, a_N'$.

Based on the normalized attribute set, the clustering threshold ε is determined as

$$\varepsilon = \frac{\max(\widetilde{a}') - \min(\widetilde{a}')}{N}. \tag{5.2}$$

The minimum number of neighbors required to form a cluster, denoted as $MinPts$, specifies the number of attribute points within a neighborhood that are grouped into a partitioned region. This approach enables effective binning of continuous attributes, while maintaining computational efficiency and robustness.

5.3.2 *Tabular Data Generation*

In this module, we employ a generative adversarial model to simulate latent data, thereby generating synthetic tabular data in the initial modality. Generative adversarial networks (GANs) [40] consist of a generator C and a discriminator D, engaged in a two-player minimax game. The generator aims to map samples from a latent distribution to data that closely resembles the real data distribution, while the discriminator seeks to differentiate between real and generated samples. Formally, the objective is expressed as

$$\min_C \max_D V(D, C) = \mathbb{E}_{x \sim p(x)}[\log D(x)] + \mathbb{E}_{z \sim p_z(z)}[\log(1 - D(C(z)))], \tag{5.3}$$

where $p(x)$ represents the true data distribution, and $p_z(z)$ is typically a simple distribution, such as a standard normal distribution.

The generator implicitly defines a distribution $p_g(x)$ as the output distribution of $C(z)$, where $z \sim p_z(z)$. For a fixed generator, the optimal discriminator is given by $D(x) = \frac{p(x)}{p_g(x)+p(x)}$. Under this optimal discriminator, the adversarial training process minimizes the Jensen-Shannon divergence (JSD) between $p(x)$ and $p_g(x)$ [40]. The global equilibrium is achieved when $p_g(x)$ matches $p(x)$, indicating that the generated and real data distributions are identical.

To adapt to tabular data generation, we adopt a design similar to CTGAN proposed in [13], modeling the distribution of tabular data and sampling from the corresponding distribution. Unlike CTGAN, which employs mode-specific normalization to address the challenges posed by the non-Gaussian and multimodal distributions of tabular data attributes, our attribute binning module achieves unified aggregation and encapsulation of attributes.

In traditional GANs, the generator is fed with vectors sampled from a standard multivariate normal distribution. However, this training approach fails to address the imbalance inherent in unified binned attributes. If the training data is randomly sampled during the training process, rows belonging to minor categories may not be adequately represented, causing the generator to fail in learning these distributions effectively. Employing general strategies to handle imbalances, such as oversampling or undersampling, often leads to a mismatch between the distribution learned by the generator and the actual attribute distribution in the data.

To address this issue, we employ a conditional GAN framework. During training, conditional GAN effectively resamples all categories of discrete attributes in a uniformly balanced manner, ensuring equitable representation. At the testing stage, the original data distribution is restored. A conditional GAN integrates the generator to model the conditional distribution of rows given specific values in discrete columns. For a value k^* in a discrete column $D_{i^*} = k^*$, the generated samples must match the condition such that $\hat{\mathbf{r}} \sim \mathbb{P}_{\mathscr{C}}(\text{row}|D_{i^*} = k^*)$. Thus, the original distribution can be reconstructed as

$$\mathbb{P}(row) = \sum_{k \in D_{i^*}} \mathbb{P}_{\mathscr{C}}(row|D_{i*} = k^*)\mathbb{P}(D_{i^*} = k). \tag{5.4}$$

During the training process, the conditional generator is capable of generating arbitrary sets of one-hot encoded discrete vectors. To guide this generation, a cross-entropy penalty is applied between the real attributes and the generated attributes across all instances in the batch. Building upon the Wasserstein GAN with gradient penalty framework [41], the corresponding loss function is designed as

$$\text{L} = E_{C(z)\sim P_g}\left[D(C(z))\right] - E_{x\sim P_r}\left[D(x)\right] + +\lambda E_{\hat{x}\sim P_{\hat{x}}}\left[\left(\left\|\nabla_{\hat{x}} D(\hat{x})\right\| - 1\right)^2\right]. \tag{5.5}$$

In the loss function, the first two terms represent the original Wasserstein GAN loss components [42], while the final term corresponds to the gradient penalty loss $\hat{x} \sim P_{\hat{x}}$ [41], which is used to regulate the discriminator's gradient over random samples. Here, P_r and P_g denote the distributions of real and generated data, respectively. The parameter λ represents the gradient penalty coefficient, and $\hat{x}$ refers to samples obtained by linear interpolation between real data points x.

5.3.3 *Behavior Molecule Construction*

To capture the internal structural relationships of behaviors, we model them as behavioral molecule graphs. These graphs are represented as $\mathscr{G} = (\mathscr{V}, \varepsilon, P) = (A, P)$, where $\mathscr{V}$ denotes the set of nodes, ε the set of edges, $P \in \mathbb{R}^{N \times n}$ is the node feature matrix (with the ith row, p_i, representing the feature vector of node v_i), and A is the adjacency matrix. In this model, nodes encode the attribute information of behaviors, while edges capture the inter-attribute relationships.

The behavior molecule construction module computes the edge structure based on the distribution of attribute pairs across the dataset. Let X represent the behavioral dataset, with each behavior record described by multiple attributes. Specifically, the kth record in X is denoted as $x_k = (x_k^1, x_k^2, \ldots, x_k^i, \ldots, x_k^n)$, where x_k^i is the attribute value at the ith position.

The behavior molecule graphs consist of the feature matrix P and the adjacency matrix A. The matrix $P = [p_1, p_2, p_3, \ldots, p_n]$ encodes the attribute information of n dimensions. The connectivity score matrix $S = (s_{ij})_{n\times n}$ represents the connectivity within the behavior molecule graph. The values of P and S are computed as follows:

$$\begin{cases} p_i = x_k^i, \\ s_{ij} = \dfrac{N_{ij}}{N_i + N_j}, \quad i, j \in \mathcal{N}, \end{cases} \tag{5.6}$$

where $\mathcal{N} = 1, 2, \ldots, n$ represents the attribute dimensions, x_k^i is the attribute of the behavior record x_k, N_i is the frequency of x_k^i in the dataset, and N_{ij} is the frequency of co-occurrence of x_k^i and x_k^j within a record. The connectivity score s_{ij} represents the co-occurrence ratio, which reflects the association between attributes x_k^i and x_k^j.

Using the connectivity score matrix S, we construct the adjacency matrix $A = (n_{ij})_{n\times n}$ for each behavior molecule graph. To control the graph structure, we introduce a connectivity score threshold, c. Specifically, if $s_{ij} < c$, then $n_{ij} = 0$ (no edge), and if $s_{ij} \geq c$, then $n_{ij} = 1$ (edge exists).

The score s_{ij} quantifies the relative co-occurrence of attributes x_k^i and x_k^j, serving as a measure of their potential association. If $s_{ij} < c$, the attributes are considered uncorrelated, and no edge is formed between them in the behavior graph. Conversely, if $s_{ij} \geq c$, an edge is created, indicating an association between the attributes. This binary edge structure effectively distinguishes attribute relationships within the graph.

After applying this process to all behavior records in X, we obtain a set of behavior molecule graphs $\widetilde{\mathcal{G}} = \{\mathcal{G}_1, \mathcal{G}_2, \ldots, \mathcal{G}_N\}$, where each graph corresponds to a specific behavioral record. Each behavior graph has n nodes, corresponding to the number of attributes in the record. Finally, we utilize graph neural network for graph-level learning on the collection of $\widetilde{\mathcal{G}}$.

5.4 Evaluation of TabBMG

In this section, we conduct a comprehensive evaluation of TabBMG's effectiveness in real-world tabular data synthesis tasks. To assess the high quality of the synthetic data generated by TabBMG, we compare it against data synthesized using CTGAN, a baseline method for tabular data generation in the original modality. The evaluation involves training a variety of machine learning models on the synthetic datasets and comparing the resulting performance. To ensure a robust and reliable assessment, we perform evaluations across multiple dimensions and metrics. To evaluate the performance of the method, we utilize three commonly used metrics: accuracy (Acc), recall (Rec), and F1 score (F1 score).

In our TabBMG framework, we utilize GIN [43] as the graph-level learning module. To compare the performance of our approach, we train models on table data synthesized by CTGAN using a range of widely used machine learning methods as well as state-of-the-art deep learning techniques. These methods include KNN, two Bayesian algorithms (GaussianNB and BernoulliNB), two decision tree algorithms (DecisionTreeGini and DecisionTreeEntropy), SVM [44], random forest [45], XGBoost [46], CatBoost [47], LightGBM [48], and the neural network-based approach DANet [49].

5.4.1 Experimental Setup

In the behavioral attribute Binning, we employ a clustering approach to partition the continuous attribute set, denoted as Π_A. This partition is then used to assign labels to each value within the attribute, enabling further computations. Specifically, we apply the DBSCAN algorithm [39], a density-based clustering method, to partition the attributes. Prior to clustering, we normalize the attributes to mitigate the influence of extreme values. Let $\mathbb{X} = x_1, x_2, \ldots, x_n$ represent an attribute. To perform mean normalization, we apply the following transformation:

$$x_i' = \frac{x_i - \overline{x}}{\sigma}, \quad i = 1, 2, \ldots, n,$$

where n is the number of data points, $\overline{x}$ is the mean, and σ is the standard deviation. The normalized attribute set is $\mathbb{X}' = x_1', x_2', \ldots, x_n'$. We then define the clustering threshold based on the normalized data as:

$$\delta = \frac{\max(\mathbb{X}') - \min(\mathbb{X}')}{n}.$$

The minimum number of nearest neighbors is set to $MinPts = 5$.

After binning the attributes, which typically results in a small label count $l_k \ll n$ for each attribute, we proceed with behavior molecule generation. During the computation of the connectivity matrix for the behavior molecule graph, we select a connectivity score threshold of $c = 0.1$. In subsequent ablation experiments, we vary the hyperparameter c to examine its impact, as it reflects the correlation between attributes. After constructing the behavior molecule graph for each data entry, we apply GIN to perform graph-level classification on the resulting behavior molecule dataset. To ensure the robustness of our results, we report the average values obtained from 10 experimental runs.

Evaluating generative models is inherently complex, as different metrics can yield significantly varying results. To ensure a comprehensive assessment, we evaluate multiple metrics across various datasets. For real-world tabular datasets, classification tasks are commonly used in machine learning. Accordingly, we assess

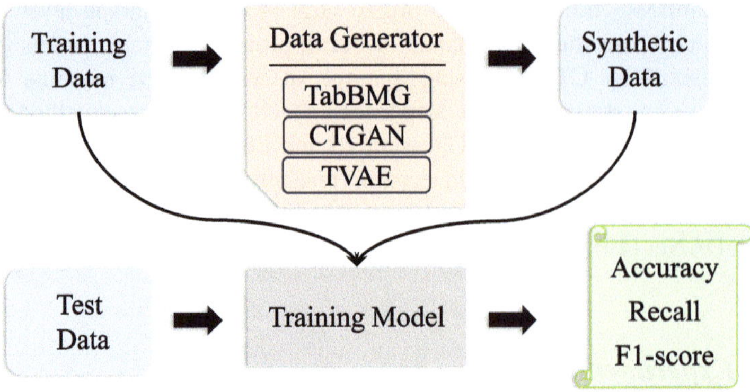

Fig. 5.2 Evaluation framework for real-world data in the experiments

the effectiveness of synthetic data generation methods based on their impact on machine learning performance. Figure 5.2 illustrates our evaluation framework.

5.4.2 Results

In this section, we compare the Acc, Rec, and F1 score obtained by training the behavior molecule data synthesized using the TabBMG method with GIN, against the baseline methods trained on table data synthesized using CTGAN and TVAE. Tables 5.1 and 5.2 summarize the average scores across real-world datasets. It is evident from the table that our method achieves the best or second-best results across all datasets.

In this experiment, we replace the tabular data generation module in the BMG framework with CTGAN and TVAE, respectively, and conduct experiments where the synthesized tabular data undergoes modality dimensionality expansion. Following the experimental design, we employ GIN for graph-level learning. We compare the two variants of the BMG framework with the original versions of tabular data synthesized using only CTGAN and TVAE. A range of widely used machine learning methods is applied to evaluate the efficiency of machine learning enhanced by synthetic data. Overall, regardless of the type of tabular data generation module used in the BMG framework, its performance in machine learning tasks—measured by metrics such as Acc, Rec, and F1 score—outperforms that of the original tabular data generation methods. This confirms our conclusion that introducing relational dependencies between attributes in the tabular data generation process, akin to the structural relationships between attributes in behavioral data, leads to higher-quality data generation, similar to the way computational drug design incorporates structural relationships in molecules. Additionally, this suggests that by binning the attributes of the tabular data, we have effectively reduced the noise at the

Table 5.1 The comparison of the machine learning efficiency after augmenting with synthetic data generated by the TabBMG method and traditional tabular data generation methods on credit approval

Dataset		Credit approval		
Metric		Acc	Rec	F1 score
C	KNN	0.680±0.024	0.665±0.030	0.665±0.027
T	GaussianNB	0.796±0.022	0.765±0.014	0.774±0.019
G	BernoulliNB	0.793±0.036	0.788±0.035	0.786±0.035
A	DecisionTreeGini	0.850±0.017	0.858±0.013	0.847±0.016
N	DecisionTreeEntropy	0.818±0.016	0.811±0.013	0.811±0.015
	SVM	0.717±0.024	0.679±0.016	0.682±0.020
	RandomForest	0.873±0.023	0.866±0.019	0.868±0.022
	XgBoost	0.855±0.015	0.848±0.011	0.849±0.014
	Catboost	0.851±0.021	0.858±0.017	0.849±0.021
	LightGBM	0.865±0.023	0.860±0.022	0.860±0.022
	DANet	0.844±0.025	0.846±0.025	0.843±0.026
T	KNN	0.669±0.036	0.662±0.039	0.661±0.037
V	GaussianNB	0.661±0.037	0.755±0.029	0.755±0.031
A	BernoulliNB	0.794±0.034	0.791±0.044	0.789±0.040
E	DecisionTreeGini	0.839±0.031	0.847±0.025	0.838±0.031
	DecisionTreeEntropy	0.811±0.028	0.812±0.030	0.809±0.029
	SVM	0.691±0.047	0.670±0.044	0.660±0.051
	RandomForest	0.660±0.051	0.660±0.051	0.660±0.051
	XgBoost	0.852±0.019	0.855±0.017	0.851±0.019
	Catboost	0.839±0.031	0.847±0.025	0.838±0.031
	LightGBM	0.848±0.032	0.849±0.034	0.846±0.033
	DANet	0.844±0.030	0.844±0.030	0.843±0.030
BMG-CTGAN		0.938±0.005	0.937±0.005	0.937±0.005
BMG-TVAE		**0.945±0.015**	**0.945±0.016**	**0.944±0.016**

attribute boundaries, thereby facilitating information aggregation. Through modality dimensionality expansion, we achieve the generation of more realistic data that better captures the complexities of real-world data.

On the other hand, we observe that the impact of using CTGAN and TVAE in the tabular data generation module of the BMG architecture varies across different datasets. This finding aligns with the results from the original modal tabular data generation approach for data augmentation in machine learning, as well as with the conclusions drawn by Xu et al. in [13]. Both variants exhibit strengths and weaknesses in performance, suggesting that the tabular data synthesis module is not the critical component of the BMG architecture. This further corroborates the significance of our information aggregation and behavior molecule construction modules.

Table 5.2 The comparison of the machine learning efficiency after augmenting with synthetic data generated by the TabBMG method and traditional tabular data generation methods on heart disease

Dataset		Heart disease		
Metric		Acc	Rec	F1 score
C	KNN	0.486±0.038	0.203±0.031	0.174±0.035
T	GaussianNB	0.522±0.058	0.315±0.058	0.290±0.059
G	BernoulliNB	0.542±0.038	0.270±0.033	0.249±0.037
A	DecisionTreeGini	0.538±0.046	0.200±0.000	0.140±0.008
N	DecisionTreeEntropy	0.486±0.044	0.294±0.037	0.289±0.029
	SVM	0.575±0.056	0.303±0.052	0.285±0.056
	RandomForest	0.573±0.053	0.305±0.048	0.287±0.041
	XgBoost	0.539±0.038	0.290±0.060	0.275±0.054
	Catboost	0.572±0.031	0.273±0.041	0.241±0.042
	LightGBM	0.241±0.042	0.262±0.036	0.242±0.042
	DANet	0.541±0.059	0.283±0.034	0.257±0.027
T	KNN	0.486±0.039	0.217±0.020	0.186±0.028
V	GaussianNB	0.517±0.077	0.337±0.051	0.300±0.043
A	BernoulliNB	0.517±0.077	0.337±0.051	0.300±0.043
E	DecisionTreeGini	0.503±0.055	0.210±0.017	0.154±0.027
	DecisionTreeEntropy	0.494±0.057	0.290±0.046	0.282±0.048
	SVM	0.494±0.057	0.290±0.046	0.282±0.048
	RandomForest	0.566±0.058	0.299±0.045	0.278±0.052
	XgBoost	0.530±0.054	0.286±0.060	0.267±0.056
	Catboost	0.566±0.069	0.289±0.037	0.267±0.052
	LightGBM	0.566±0.059	0.285±0.038	0.254±0.036
	DANet	0.531±0.063	0.303±0.041	0.271±0.039
BMG-CTGAN		**0.803±0.013**	**0.682±0.024**	**0.667±0.024**
BMG-TVAE		0.792±0.025	0.676±0.041	0.663±0.049

To visually illustrate the behavior molecule graphs constructed from real-world data and those synthesized using the BMG framework, we present examples in Fig. 5.3. The figure showcases behavior molecule graphs generated from real-world datasets alongside those constructed from synthetic datasets produced by the BMG framework. From the figure, it is evident that the graphs derived from the original data exhibit sparse edge structures and limited representational capability due to the dispersed distribution of attributes. In contrast, the high-quality behavior molecule graphs synthesized by the BMG framework feature significantly richer edge structures. This improvement is primarily attributed to the attribute binning module, the first component of the BMG architecture, which effectively reduces noise at the attribute margins and plays a pivotal role in information aggregation.

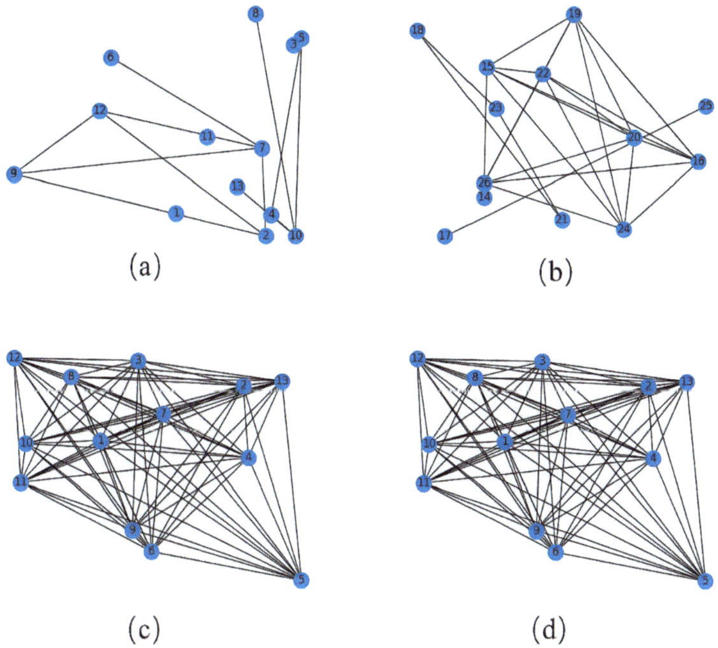

Fig. 5.3 Visualization of behavior molecule graphs corresponding to the original data from the disease dataset (**a**) and (**b**) and those generated through the BMG framework (**c**) and (**d**)

5.4.3 Parameter Analysis

A critical parameter in our BMG framework is the ratio of synthetic data to real data, a topic that has recently garnered significant attention in research [16]. Specifically, this pertains to determining the optimal proportion of real to synthetic data for enhancing machine learning performance. In this section, we conduct experimental investigations to address this question.

We conduct experiments using datasets composed of varying proportions of real and synthetic data to investigate the impact of synthetic data ratios on machine learning performance. The results of these experiments are presented in Fig. 5.4, and a detailed analysis of these findings is provided below.

On one hand, compared to using only original data without synthetic augmentation, incorporating an appropriate amount of synthetic data generally enhances machine learning performance. This improvement occurs because synthetic data partially simulates unknown data distributions, thereby enhancing the model's generalization capability and enabling it to perform better on test datasets. As the proportion of synthetic data decreases, the performance gains also gradually diminish, aligning with intuitive expectations. When the proportion of synthetic data is zero, the scenario degenerates to using only original data, resulting in no performance improvement.

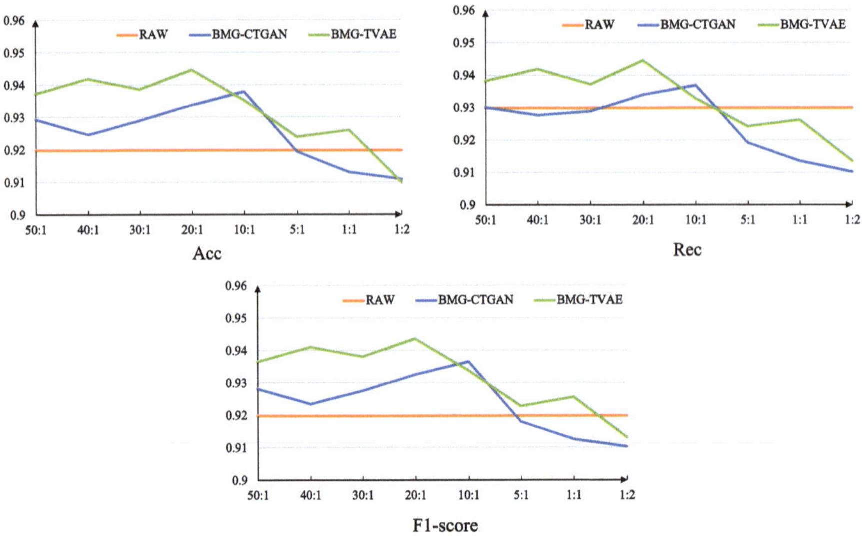

Fig. 5.4 Impact of different synthetic data proportions on machine learning performance (Acc, Rec, and F score) for the credit approval dataset

On the other hand, as the proportion of synthetic data increases, the performance gains initially rise. In our experiments, the performance boost reaches a peak when the ratio of real to synthetic data is approximately 10:1. However, further increasing the proportion of synthetic data leads to a noticeable decline in machine learning performance, eventually falling below the performance of models trained without synthetic data. This decline is attributed to the model becoming overly sensitive to synthetic data when its proportion is excessively high, thereby reducing its ability to distinguish between real and synthetic distributions. Consequently, the model's performance on test datasets deteriorates.

Overall, achieving performance enhancement in machine learning through synthetic data requires an appropriate balance between real and synthetic data. However, determining the optimal ratio for maximizing performance improvement remains an open challenge without a precise computational method. Recent studies have explored whether synthetic data contributes positively or negatively to machine learning performance [16], but a systematic approach to calculating and analyzing the optimal real-to-synthetic data ratio is yet to be established. With the rapid advancements in generative methods in recent years, we believe this topic will become a significant focus of future research.

Another critical parameter in our framework is the degree of aggregation during attribute binning, represented by $MinPts$. This parameter determines the level of aggregation applied to attributes within the behavioral data. To evaluate its impact, we conducted experiments on the credit approval dataset using various levels of aggregation, with detailed results presented in Table 5.3.

Table 5.3 The impact of varying aggregation levels (MinPts) on machine learning performance for the credit approval dataset

MinPts		2	5	8	11	14
BMG	Acc	0.8735	**0.9382**	0.9330	0.9241	0.9315
	Rec	0.8743	**0.9371**	0.9336	0.9217	0.9300
	F1 score	0.8523	**0.9373**	0.9312	0.9225	0.9301

Our observations indicate that low levels of aggregation lead to a significant decline in performance. This is attributed to the generative model being overly influenced by noise within the continuous attributes of the tabular data. Continuous attributes often contain substantial noise at the distribution margins, which, if not mitigated, can cause the generative model to produce irrelevant noise-laden synthetic data, resulting in substantial degradation of machine learning performance.

Conversely, when the aggregation level is excessively high, the noise issue in continuous attributes is effectively addressed; however, the excessive aggregation compromises attribute diversity. This reduction in diversity leads to overly uniform synthetic data, which fails to provide the model with richer latent information. Consequently, the machine learning model becomes prone to overfitting known information, ultimately leading to a decline in performance.

In summary, achieving optimal machine learning performance requires a balanced approach. It is essential to sufficiently attenuate the noise in continuous attributes, while preserving a certain level of attribute diversity. This balance ensures the generation of more informative and diverse synthetic data, which can effectively enhance machine learning outcomes.

5.5 Conclusion

This chapter presents a novel tabular data generation framework, TabBMG, which integrates the relationships between attributes in tabular data through dimensional elevation during the data generation process. The TabBMG framework aims to reduce edge noise in continuous attributes of tabular data through attribute binning and subsequently generate high-quality synthetic data by leveraging the latent relationships among internal attributes within the dataset.

Our study demonstrates that synthetic data generated under the TabBMG framework achieves optimal enhancement of machine learning performance. Furthermore, through parameterized experiments, we address a critical issue in current research: whether synthetic data positively impacts machine learning and how the ratio of real-world data to synthetic data influences learning efficiency. Our results indicate that a proper balance between synthetic and real-world data is essential.

When the proportion of synthetic data is small, its enhancement effect on machine learning is minimal, closely resembling the performance of models trained solely on real-world data. As the proportion of synthetic data increases to an optimal level, it effectively simulates unknown potential data, enhancing the generalization capability of machine learning models and significantly improving their performance on test datasets. However, this enhancement is not unlimited. When the proportion of synthetic data surpasses that of real-world data, machine learning models tend to overfit the synthetic data, reducing sensitivity to real-world data, which ultimately results in a marked decline in learning efficiency. Thus, achieving an appropriate ratio of synthetic to real-world data is critical for maximizing the benefits of synthetic data in machine learning.

Additionally, we analyzed the effect of the aggregation parameter (MinPts) in the attribute binning module. When the level of aggregation is too low, edge noise in continuous attributes of tabular data significantly affects the generation process. Overfitting to such noise causes generative models to produce irrelevant or noisy data, thereby severely degrading machine learning performance. Conversely, excessive aggregation diminishes the richness of the data, resulting in synthetic data that lacks diversity and fails to provide sufficient simulated information, similarly leading to a decline in performance.

In summary, generating high-quality synthetic tabular data to enhance machine learning efficiency requires not only an optimal balance between synthetic and real-world data but also careful handling of attribute characteristics within the data. While significant progress has been made in synthetic data generation methods, limited research has been conducted on the impact of the synthetic-to-real data ratio on learning performance. We propose to further investigate this issue, which will advance the practical applications of synthetic data in machine learning.

References

1. S. Hegselmann, A. Buendia, H. Lang, M. Agrawal, X. Jiang, D. Sontag, in *International Conference on Artificial Intelligence and Statistics* (PMLR, New York, 2023), pp. 5549–5581
2. V. Borisov, T. Leemann, K. Seßler, J. Haug, M. Pawelczyk, G. Kasneci, IEEE Trans. Neural Networks Learn. Syst. **35**, 7499–7519 (2022)
3. A. Jain, H. Patel, L. Nagalapatti, N. Gupta, S. Mehta, S. Guttula, S. Mujumdar, S. Afzal, R. Sharma Mittal, V. Munigala, in *Proceedings of the 26th ACM SIGKDD International Conference on Knowledge Discovery & Data Mining* (2020), pp. 3561–3562
4. M. Bansal, A. Goyal, A. Choudhary, Decision Analytics Journal **3**, 100071 (2022)
5. W. Bialek, Proc. Natl. Acad. Sci. **119**(18), e2021860119 (2022)
6. Y. Lu, M. Shen, H. Wang, X. Wang, C. van Rechem, T. Fu, W. Wei, arXiv preprint arXiv:2302.04062 (2023)
7. B. Liu, M. Ding, S. Shaham, W. Rahayu, F. Farokhi, Z. Lin, ACM Comput. Surv. (CSUR) **54**(2), 1 (2021)
8. S.Z. El Mestari, G. Lenzini, H. Demirci, Comput. Secur. **137**, 103605 (2024)
9. Q. Qu, Y. Shen, X. Chen, Y.Y. Chung, T. Liu, in *Proceedings of the AAAI Conference on Artificial Intelligence*, vol. 38 (2024), pp. 4632–4640
10. T. Hu, J. Zhang, R. Yi, Y. Du, X. Chen, L. Liu, Y. Wang, C. Wang, in *Proceedings of the AAAI Conference on Artificial Intelligence*, vol. 38 (2024), pp. 8526–8534
11. Y. Li, K. Zhou, W.X. Zhao, J.R. Wen, arXiv preprint arXiv:2303.06574 (2023)

12. M. Gulati, P. Roysdon, in *Advances in Neural Information Processing Systems*, vol. 36, pp. 7333–7343 (2024)
13. L. Xu, M. Skoularidou, A. Cuesta-Infante, K. Veeramachaneni, in *Advances in Neural Information Processing Systems*, vol. 32, pp. 7333–7343 (2019)
14. A. Kotelnikov, D. Baranchuk, I. Rubachev, A. Babenko, in *International Conference on Machine Learning* (PMLR, New York, 2023), pp. 17,564–17,579
15. L. Hansen, N. Seedat, M. van der Schaar, A. Petrovic, Adv. Neural Inf. Proces. Syst. **36**, 33781 (2023)
16. Y. Wang, J. Zhang, Y. Wang, in *The Twelfth International Conference on Learning Representations, ICLR 2024, Vienna, Austria, May 7–11, 2024* (OpenReview.net, New York, 2024)
17. Y. Lin, T. Ruan, J. Liu, H. Wang, IEEE Trans. Knowl. Data Eng. **36**, 1431–1449 (2023)
18. X. Guo, L. Zhao, IEEE Trans. Pattern Anal. Mach. Intell. **45**(5), 5370 (2022)
19. D.S. Wigh, J.M. Goodman, A.A. Lapkin, Wiley Interdiscip. Rev.: Comput. Mol. Sci. **12**(5), e1603 (2022)
20. X. Zeng, H. Xiang, L. Yu, J. Wang, K. Li, R. Nussinov, F. Cheng, Nat. Mach. Intell. **4**(11), 1004 (2022)
21. D. Weininger, J. Chem. Inf. Comput. Sci. **28**(1), 31 (1988)
22. Z. Li, M. Jiang, S. Wang, S. Zhang, Drug Discov. Today **27**(12), 103373 (2022)
23. M. Ma, X. Lei, Comput. Biol. Med. **169**, 107911 (2024)
24. K. Atz, F. Grisoni, G. Schneider, Nat. Mach. Intell. **3**(12), 1023 (2021)
25. Y. Wang, J. Wang, Z. Cao, A. Barati Farimani, Nat. Mach. Intell. **4**(3), 279 (2022)
26. Z. Xiong, D. Wang, X. Liu, F. Zhong, X. Wan, X. Li, Z. Li, X. Luo, K. Chen, H. Jiang, et al., J. Med. Chem. **63**(16), 8749 (2019)
27. C. Shang, Q. Liu, Q. Tong, J. Sun, M. Song, J. Bi, Neurocomputing **445**, 12 (2021)
28. C. Lee, J. Kim, N. Park, in *International Conference on Machine Learning* (PMLR, New York, 2023), pp. 18940–18956
29. Z. Zhao, A. Kunar, R. Birke, L.Y. Chen, in *Asian Conference on Machine Learning* (PMLR, New York, 2021), pp. 97–112
30. J.P. Reiter, J. Off. Stat. **21**(3), 441 (2005)
31. L. Aviñó, M. Ruffini, R. Gavaldà, arXiv preprint arXiv:1807.01514 (2018)
32. J. Zhang, G. Cormode, C.M. Procopiuc, D. Srivastava, X. Xiao, ACM Trans. Database Syst. (TODS) **42**(4), 1 (2017)
33. N. Patki, R. Wedge, K. Veeramachaneni, in *2016 IEEE International Conference on Data Science and Advanced Analytics (DSAA)* (IEEE, New York, 2016), pp. 399–410
34. Y. Sun, A. Cuesta-Infante, K. Veeramachaneni, in *Proceedings of the AAAI Conference on Artificial Intelligence*, vol. 33 (2019), pp. 5049–5057
35. B. Wen, Y. Cao, F. Yang, K. Subbalakshmi, R. Chandramouli, in *ICLR Workshop on Deep Generative Models for Highly Structured Data* (2022)
36. J. Kim, J. Jeon, J. Lee, J. Hyeong, N. Park, in *Proceedings of the Web Conference 2021* (2021), pp. 1506–1515
37. M. Cvitkovic, arXiv preprint arXiv:2002.02046 (2020)
38. C.T. Li, Y.C. Tsai, J.C. Liao, in *2023 IEEE 39th International Conference on Data Engineering (ICDE)* (IEEE, New York, 2023), pp. 3589–3592
39. M. Ester, H.P. Kriegel, J. Sander, X. Xu, et al., in *KDD*, vol. 96 (1996), pp. 226–231
40. I. Goodfellow, J. Pouget-Abadie, M. Mirza, B. Xu, D. Warde-Farley, S. Ozair, A. Courville, Y. Bengio, in *Advances in Neural Information Processing Systems*, vol. 27, pp. 2672–2680 (2014)
41. I. Gulrajani, F. Ahmed, M. Arjovsky, V. Dumoulin, A.C. Courville, in *Advances in Neural Information Processing Systems*, vol. 30, pp. 5767–5777 (2017)
42. S.C.M. Arjovsky, L. Bottou, in *arXiv:1701.07875* (2017)
43. K. Xu, W. Hu, J. Leskovec, S. Jegelka, in *Proceedings of the ICLR, New Orleans, LA, USA, May 2019* (2019)
44. C. Cortes, V. Vapnik, Mach. Learn. **20**, 273 (1995)
45. L. Breiman, Mach. Learn. **45**, 5 (2001)

46. T. Chen, C. Guestrin, in *Proceedings of the ACM SIGKDD, San Francisco, CA, USA, August 2016* (2016), pp. 785–794
47. A.V. Dorogush, V. Ershov, A. Gulin, arXiv preprint arXiv:1810.11363 (2018)
48. G. Ke, Q. Meng, T. Finley, T. Wang, W. Chen, W. Ma, Q. Ye, T.Y. Liu, in *Advances in Neural Information Processing Systems*, vol. 30, pp. 3146–3154 (2017)
49. J. Chen, K. Liao, Y. Wan, D.Z. Chen, J. Wu, in *Proceedings of the AAAI, March 2022*, vol. 36 (2022), pp. 3930–3938

Chapter 6
Identifying Hidden Cybercrime Behavior by Behavior Tomographer

Abstract Identifying hidden cybercrimes is a challenging task, as these behaviors are often carefully planned by criminals with counter-surveillance awareness. Existing solutions for cybercrime detection struggle to uncover enough clues to identify hidden criminal behaviors. Malicious behaviors are concealed beneath benign behaviors, and the boundaries between malicious and benign behaviors in the representation space are blurred to evade mainstream deep learning-based security authentication models. This work introduces a *b*ehavior *t*omographer (BT) to reconstruct the behavior structure from three slices: agent, event, and attribute slices, enabling more granular detection of hidden cybercrimes. The core idea of BT is to reconstruct interior information about behavior structure from multiple slices, much like computed tomography in modern medicine enables the reconstruction of internal body. It enables the extraction of discriminative information from intricate interior associations between behavioral attributes rather than surface information meticulously crafted by criminals. Our experiments are conducted on two representative cybercrime datasets. Promising experimental results demonstrate that BT outperforms state-of-the-art models on key metrics, achieving around 0.99 AUC-ROC and approximately 0.9 AUC-PR. Moreover, BT notably excels at low false positive rates, showcasing its high effectiveness for real-world applications.

6.1 Introduction to Topological Reconstruction

Motivation As criminal behaviors continue to proliferate, security professionals have developed intricate solutions to thwart a deluge of illegal activities [1, 2]. To evade detection, criminals gradually abandon overt methods and conceal their obvious criminal features during illegal activities. They employ technologies such as anonymization services to technically conceal their identities and mimic the behavioral patterns of ordinary users during the commission of their crimes. For example, online brushing fraud has evolved into one of the most prevalent and rapidly mutating forms of financial fraud in China [3]. It has even become an entry point for more complex composite frauds involving fake investment and lending, as well as other cybercrimes. In typical cases, victims are lured into becoming

C. Wang, *Computational Structural Behavior*,
https://doi.org/10.1007/978-981-95-1013-9_6

"promotional agents" through promises of small commissions for completing online brushing tasks. The criminal actors then exploit the victims' greed by encouraging them to deposit larger sums, claiming that higher investments will yield greater rebates. However, the victims' funds are ultimately siphoned into the criminals' own bank accounts, with no genuine rebates or payouts materialized. Online brushing fraud exhibits a high degree of concealment. The perpetrators carry out the entire fraud process through purely online operations, without any physical contact with the victims. Furthermore, they often acquire a comprehensive set of stolen personal information, including bank cards, social media accounts, and payment IDs, prior to committing the fraud. This makes it extremely difficult for law enforcement to obtain real personal information or establish effective connections with criminals.

Hidden cybercrimes are challenging to identify, exposing ordinary Internet users to risks and harms such as financial losses, data breaches, and service disruptions [4–6]. Therefore, accurate detection of these criminal behaviors is crucial in mitigating widespread damages across various domains. To effectively prevent and respond to these cybercrimes, security analysts deploy various tools to record, track, and inspect web behaviors. They even employ powerful deep learning algorithms to identify hidden cybercrimes.

Existing security solutions [7, 8] strive to identify hidden criminal behaviors by projecting behaviors in a well-separated vector space and then exploring the associations of behaviors. They regard criminal behaviors as the rare ones that deviate from known behavioral patterns, namely a model that is fitted from the majority of normal recorded behaviors [9, 10]. Essentially, these solutions distort and deform the original input of behaviors through affine and nonlinear transformations until different behaviors can be easily separated from others. They first map events into vectors in vector space and then divide events into different regions by considering the distance between vectors (Sol. V in Fig. 6.1). According to Sol. V consisting of three steps, one vector corresponds to one event in the vector space. The solutions by Sol. V focus on the associations between different events and introduce metrics related to the labels of events.

Recently, many research communities focus on identifying malicious behaviors using anomaly detection solutions [11, 12]. For instance, inspired by the success of

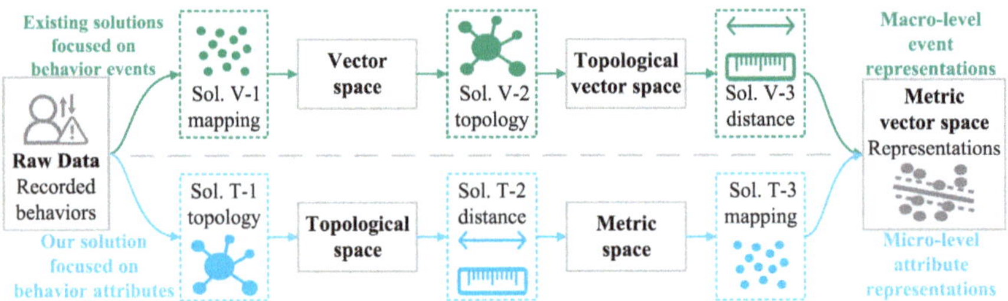

Fig. 6.1 Vector space of behavior events (Sol. V) vs. topological space of behavioral attributes (Sol. T)

deep methods, the methods based on autoencoders [13, 14] or generative adversarial networks [7, 15] have been extensively studied to learn the representations that capture the distribution of normal patterns. They decide whether a data object is normal or deviates too much from normality by using reconstruction errors in an intermediate representation space where normal and abnormal data objects are supposedly well separated. Although current research has achieved some success, it is worth noting that existing data-driven security solutions have performed well by leveraging surface associations between behaviors. These surface associations often derive from operational sequences or shared characteristics between behaviors. In most cases, researchers represent surface associations between behaviors in intuitive forms, such as sequences and graphs.

To sum up, the key challenge lies in the fact that existing models rely solely on coarse-grained representations of behavior, failing to gather sufficient clues to identify hidden criminal behaviors. When criminals intentionally hide their illicit actions, the intuitive correlations become less apparent, and valuable information is often deliberately obfuscated by wrongdoers. As a result, the boundaries between malicious and benign behaviors in the representation space become blurred.

Solution Behavioral data with even a small number of attributes could encode a high-dimensional structure representing the connectivity/associations of behavioral attributes [16–18]. Utilizing the attribute-level associations in behavior can see what lies interiorly for behavior models [19, 20] to detect criminal behaviors. As illustrated in Fig. 6.1, we can introduce a topological space based on the definition via open sets (Sol. T-1). In the topological space, the topology is on the set (X) of the fine-scale behavioral attributes and is a collection of subsets of X.[1] Considering the co-occurrences of any two attributes, we introduce metrics related to the co-occurrences to define similarity in the metric space (Sol. T-2). Next, we can represent the interior associations among all behavioral attributes as the complex network and map them to metric vector space with the help of network representation learning (Sol. T-3). The solutions by Sol. T focus on the associations between different attributes of behavior events in the topological space and introduce metrics related to the co-occurrence of attributes. Different from Sol. V, one vector corresponds to one attribute after mapping the attribute into the metric vector space.

This chapter proposes a novel method, named *b*ehavior *t*omographer (BT) that adopts fine-grained detectors to reflect interior associations within behaviors. By introducing a reformatted graph, namely behavior slice, on different associations, it images the behavior structure. In particular, BT reconstructs behavior information from different behavior slices, namely agent, event, and attribute slices (as shown in Fig. 6.2). When we focus on the associations of behaviors, the associations will be reflected with different characteristics depending on what we focus on

[1] Examples: Given $X = \{1, 2, 3, 4\}$ consisting of four attributes, the family $\{\emptyset, \{1, 2\}, \{2, 3\}, \{1, 4\}, X\}$ of five subsets of X forms the topology of X. The subset of two attributes corresponds to the recorded behaviors containing two attributes.

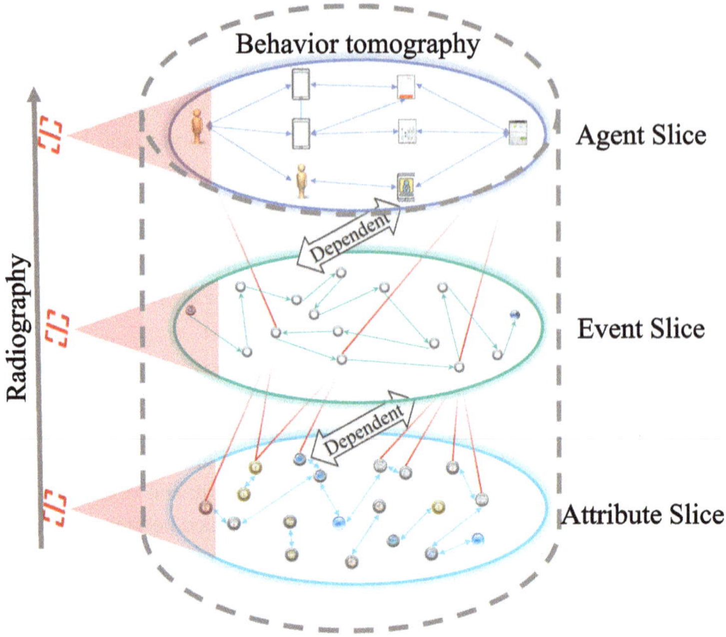

Fig. 6.2 Illustrations of behavior tomography

(e.g., the agent that performs behavior, the attribute of behavior). The associations inevitably influence each other, making it difficult to analyze specific associations in a targeted manner from noisy associations. BT independently learns multiple slice representations and gradually aggregates behavior information from attributes to the agents that reflect behavioral patterns. This eliminates the overlaying of information outside the focused associations. In BT, we design representation learning for individual behavior slices and information fusion mechanisms for different behavior slices. Ultimately, BT achieves the detection of malicious cybercrime by assessing the consistency between behavior events and behavior agents through cascading the learned representations based on agent slice and event slice.

Contributions The major contributions of this work include:

- To the best of our knowledge, this is the first work that reconstructs the associations within behavior and forms the behavioral structure to enable precise diagnosis of hidden cybercrimes.
- It proposes a novel method, behavior tomographer (BT), which can layer behavioral data into multiple slices with fine resolution and learn efficient representations on the reformatted slices by shielding irrelevant information.
- It conducts extensive experiments on representative cybercrime scenarios. They demonstrate that BT significantly outperforms the competing methods in terms of both the area under the receiver curve and the precision-recall curve. Moreover,

BT obtains a substantially better performance under the low false positive rate than baselines, which is of high importance in real-world applications.

6.2 Related Work on Identifying Cybercrime Behavior

Researchers have long focused on and modeled some critical tasks related to cybercrime to address the increasing network threats [21–24]. Cybercrime can roughly be divided into two major categories: application security (such as financial fraud, fake news, and spam comments) and network infrastructure security (such as domain name system, network traffic, and routers). We outline these common security tasks in Table 6.1.

Early explorations focused on rule-based methods [25, 26]. However, the advent of online services has opened the data-driven era, which makes the rule-based methods relying on expert experience usually weak [19, 27, 28]. Machine learning-based methods have been widely applied in cybersecurity applications to combat cybercrime. Typically, the security issues can be modeled as anomaly detection tasks to identify rare behaviors in data samples [5, 29]. Early machine learning solutions mostly utilized manually crafted application-specific features as inputs, followed by training models (such as support vector machine and ensemble-based methods [30–33]) to detect security threats. Linear-based methods [30] try to separate abnormal data objects from the others, while they may not be effective when the data are irrelevant but highly aggregated in some areas. Near neighbor-based methods [33] generally rely on the assumption that normal data objects occur near their closest data objects, while abnormal data objects will occur far from most data objects. Partition-based methods [31] define the anomalies as outliers that are more likely to be separated and circularly split the data space with a random hyperplane. Further, ensemble-based methods [32] combine the outputs of multiple base detectors and integrate these base detectors to explore subsets of different dimensions, which obtains more accurate and diverse models. Classical solutions can use existing

Table 6.1 Overview of applications to cybersecurity

Cybersecurity task		Target cybercrimes	References
Application security	Transaction security	Money laundering, cash-out, underground market activities, insurance fraud, loan default	[20, 26, 40, 49]
	Cognition security	Web spam, review spam, fake news, fake account	[35, 37, 41, 43]
Network infrastructure security	Network security	Botnet, malicious domain, intrusion detection	[10, 13, 18, 50]
	System security	Malware, system vulnerability, smart contract vulnerability	[5, 11, 28, 51]

classifiers for detecting anomalies [34]. Obviously, they are inefficient in capturing disguised behaviors (e.g., multiple consecutive pings from the same IP address, where each individual event may be deemed normal when considered in isolation) with the help of the macrolevel features of the behaviors.

Deep learning has emerged as an effective approach in the field of network security by automatically learning advanced network attack features and exhibiting considerable learning capabilities [35–37]. Mirsky et al. proposed an anomaly detection method for intrusion detection based on an ensemble of autoencoders, which are trained to reconstruct the input data [13], where the data are supposedly well separated in a lower hidden dimensional embedding space [38, 39]. They capture the relationships between different events to a certain extent, but they still take no account of the fine-scale associations between events. To improve the ability to identify hidden behaviors and the supervision of known behavior patterns, some recent studies turn to semi-supervised algorithms, which use labeled data for improving the model's performance and leverage unlabeled data to facilitate behavior modeling [34].

Some researchers have focused on the many explicit or implicit associations among cybercrimes, developing a series of sequence- and graph-based works to better detect cybercrimes [18, 20, 40]. Graph-based methods provide a new way to achieve competitive results [41–43]. They incorporate intricate dependencies into an information propagation process to learn application-relevant representations, resulting in achieving a more reliable behavioral model [44]. Network representation learning has emerged as a successful approach for identification tasks by aggregating the representations of neighbors' information in graph-structured data [40]. Velickovic et al. [45] proposed to use the attention mechanism to sum the features of neighboring nodes, which extracts more effective representations. Relative applications [46] further verify the effectiveness of graph neural networks in learning the intricate relations between data. Zeng et al. [47] abstracted high-level behaviors from audit logs by aggregating contextual semantics from different behaviors. The graph-based approaches tend to view behavior as nodes in the graph structure and further regard the detection of criminal behavior as node classifications. However, these graph-based approaches focus on modeling the behavior nodes at the macrolevel and hence face limitations in identifying the behaviors that only slightly deviate from the normal behavioral pattern. In previous works [48], a graph-based method was proposed to mine deeply the property-level (as an enhancement to the event-level) associations in behavioral data. It transforms the intensity of behavioral anomaly into the spatial decentrality of its behavioral properties. Here, we project behaviors into different slices to mask the noise caused by the superposition of information and learn sliced representations by exploiting various behavioral associations in a fine resolution. Our method can achieve expressive behavioral patterns by aggregating information from attributes to agents and perceiving behavior from attribute representations.

6.3 Overview

6.3.1 Security Model

6.3.1.1 Behavioral Data

BT focuses on structured behavioral data recorded by information systems, where behavioral data is typically collected according to predefined fields set by different business requirements, such as transaction fields in online payment systems, traffic statistics data on gateway ports, and operation records on computer operating systems. Our work explores the interior structure of behavior, aiming to detect inconsistencies between normal and criminal behaviors, with a focus on identifying criminal behaviors that deviate from specific or majority user normal behavior patterns. A behavior in this context can be seen as an event performed by an agent, composed of various behavioral attributes such as time, location, and type. Specifically, we design three types of entities from behavioral data: agent (which performs specific behavior), event (target object to be detected), and attribute (properties of the behavior). In contrast to existing works that project behaviors into a unified space (treating behaviors as vectors and representing them, e.g., using encoder-based representation learning modules), BT decomposes behaviors into finer granularity, focusing on the associations between different entities within the behavior.

6.3.1.2 Behavior Slice

Graph is an effective way to express associations, e.g., the co-occurrence relations, in many real-world settings [9, 20, 46]. In behavior tomography, we design behavior slices to describe the forms of associations within behavior. Each slice (e.g., agent slice, event slice, attribute slice) delineates a graph structure defined on different focused entities (e.g., agent, event, attribute). The three different slices in BT have distinct purposes and target-specific aspects of behavior analysis. Attribute slice aims to learn representations at the attribute level by capturing the associations among different attributes. Looking in the behavior, we focus on the topology of the behavioral attribute set and exploit the intra-associations within each behavior and the interassociations of overall behaviors in the attribute slice. Event slice concentrates on modeling the short-term historical interaction between behaviors in chronological order. Anonymous event ordering plays a positive role in modeling agents' preferences [46]. Agent slice focuses on representing agents in a low-dimensional space, in which two representations with high similarity indicate that their corresponding agents have similar preferences. Looking beyond the limitations of a single behavior, we utilize the interaction between behaviors and the information from agents with similar behavioral patterns. The agent information helps behavioral models identify criminal behaviors effectively.

6.3.1.3 Slice Representation

We design the graph-based model to learn the sliced representations on different slices independently. It does not solely rely on representing raw behavioral data, which can be easily manipulated by criminals. Instead, it delves into the interior association within behavior to obtain highly expressive representations. Along this way, we propose the intra-slice representation fusion that learns sliced representations (i.e., attribute, event, agent representations), respectively. Attribute slice models inter/intra-associations among attributes, event slice models the historical interaction between events, and agent slice utilizes the information from agents with similar behavioral patterns. Furthermore, we adopt a slice-by-slice structure to learn the agent representations by a procedure from *attribute* to *event* to *agent*, so as to better the behavioral model for identifying hidden cybercrimes insightfully. Specifically, we design the sliced representation fusion that contains: (i) aggregating event representations from microlevel attribute representations; (ii) updating event representations with sequence information; and (iii) aggregating agent representations from historical event representations that belong to corresponding agents.

6.3.1.4 Identifying Cybercrimes

BT can detect hidden cybercrimes based on the learned representations in the agent and event slices with a concatenation mechanism.

6.3.2 *Threat Model*

We consider hidden cybercrimes as criminal behaviors carried out by malicious actors using means such as stealth techniques and orchestrated attack strategies. The primary characteristic of these behaviors is to minimize the exposure of malicious intent, making the malicious behavior closely resemble benign behavior in order to evade detection. At the network level, attackers may employ various techniques to conceal their true identities. These techniques include the use of virtual private networks (VPNs), proxy servers, Tor network, etc., to alter their IP addresses and obfuscate the source and destination of network traffic, making their criminal activities appear more like normal network communications. Besides, attackers tend to create multiple fake accounts and identities to scatter the traces of fraudulent activities. They may use different devices and user agents to avoid being detected as associated with the same person or device. At the service level, attackers attempt to mimic the behavior patterns of legitimate users, including clicks, browsing, and software operations. They randomize their behavior patterns to make them appear like those of genuine users, thus avoiding detection of anomalous behavior. Additionally, they limit the frequency and quantity of malicious behaviors. Attackers avoid conducting a large number of malicious behaviors within a short period and instead spread them out over different time intervals to emulate benign behaviors.

6.4 Slice-Based Behavior Tomographer

We present the framework of behavior tomographer (BT) in Fig. 6.3. BT consists of four phases: (i) lookup embedding, which generates the primitive representations; (ii) intra-slice representation fusion, which learns the representations of all nodes encoding the intricate interactions in each slice respectively; (iii) sliced representation fusion, which learns the agent representations from fine-scale attributes to agents; (iv) prediction, which produces the result by leveraging the learned event and agent representation.

6.4.1 *Lookup Embedding*

In real applications, the recorded behaviors enable analysts to gain insight into cybersecurity where each behavior can record a lot of information. Let $B = \{b_1, b_2, \cdots, b_i, \cdots, b_{|B|}\}$ denote a set of behaviors, where $|B|$ is the cardinality of the set. Each behavior b_i has m attributes, such as Event_category, Event_place, Event_time, etc. A behavior is represented by $b_i = \{b_i^1, b_i^2, \cdots, b_i^m\}$, where b_i^j is the attribute value of the jth attribute. To model normal behavior patterns of behavioral agents, we further denote the performer of a behavior as the agent g. Our task is to identify whether a behavior is normal or not according to the normal behavior pattern of agent.

We initialize the representations of all agents, events, and attributes. In reality, the value of the agent, event, or attribute can be categorized into two types: discrete values and continuous values. For instance, the identification number on an

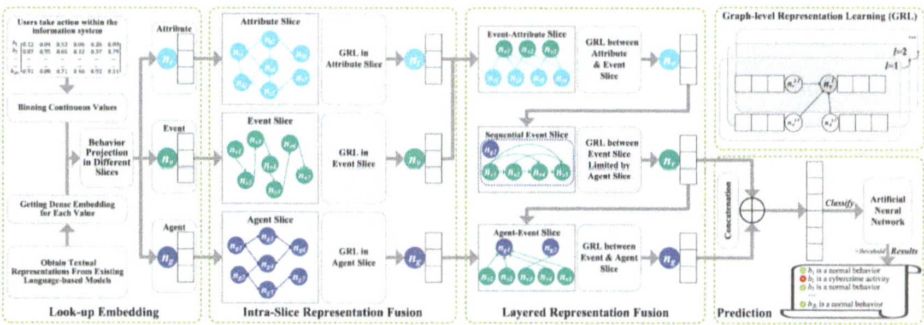

Fig. 6.3 The framework of the proposed behavior tomographer for identifying hidden cybercrimes

identification card can be regarded as a discrete identifier. In contrast, the purchase amount in shopping behavior is typically a continuous value. For continuous values, we first transform them into discrete values by using binning techniques. Then, we adopt the lookup embedding to obtain dense embedding for each value. We formulate the lookup table as $\Phi \in \mathbb{R}^{d'}$, where d' is the dimension of the embedding vectors. We term the representation of the value n as $\vec{n} = \Phi[n]$, where $\Phi[n]$ denotes the row corresponding to value n in Φ. Lookup embedding is implemented by creating a lookup table that maps each value to a corresponding dense vector representation where the table is initially populated with pretrained or randomly initialized vectors. Preferably, if existing language-based models can be utilized to obtain textual representations from the data itself or external domain knowledge, we choose these available textual representations to assist subsequent model learning. Leveraging representations encoding meaningful context helps the model learn more effectively from limited data as compared to random vectors that lack interpretability.

6.4.2 *Intra-Slice Representation Fusion*

In this part, we layer behavioral data into three different slices, i.e., *attribute slice*, *event slice*, and *agent slice*, based on different aspects of relations for learning the latent representations.

Attribute Slice Existing methods mainly focus on concatenating or aggregating the initial representations of attributes to extract the internal information of each behavior. However, the existing methods are not effective because they lose sight of the inter-level context information, i.e., the pair-wise attributes over different behaviors. Generally, a nonempty set can specify many different topologies. We simplify generating the behavior topology to how to generate the appropriate subsets of cardinality 2. So each behavior can be converted into a completed graph for learning the node representations of attributes via network representation learning.

For each b_i in the set B of all behaviors, we employ a graph $\mathcal{G}_t = (\mathcal{N}_t, \mathcal{E}_t, \mathcal{R}_t)$ to present the attribute slice, where $\mathcal{N}_t = \{b_i^j | b_i \in B, b_i^j \in b_i\}$ denotes the set of attribute values, $\mathcal{R}_t$ denotes the set of relations between the corresponding two attribute values, and $\mathcal{E}_t = \{(b_i^j, b_i^k, r) | b_i \in B\}$ denotes the edge set. Optionally, we can use information gain to select the fields in the raw data that have the highest impact on detecting cybercrimes as behavioral attributes. Each edge indicates two co-occurrence attribute values with relation $r \in \mathcal{R}_t$ in a behavior, and relation r depends on the types of attribute b_i^j and b_i^k. For instance, a "time-location" relation of edge can be generated between the attribute representing the occurrence time of a behavior and the attribute representing the location of the behavior. Figure 6.4 shows the example of an attribute slice with three behaviors. The attribute slice integrates the inter-level context by linking all pairs of attributes over all behaviors.

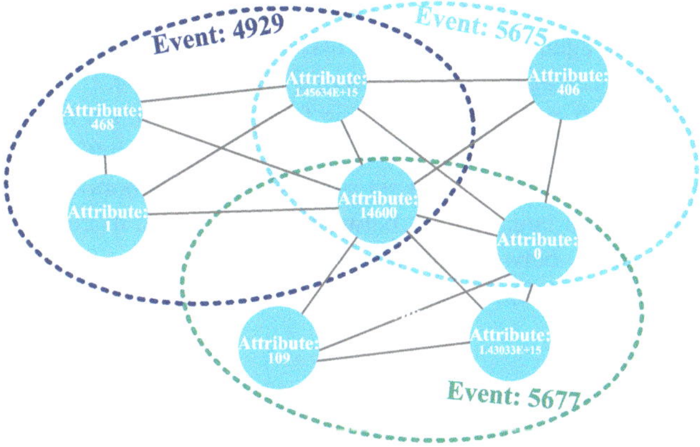

Fig. 6.4 Illustrations of the attribute slice

In the attribute slice, the relationship between any two attribute values is not only directly affected by the behavior events that contain these attribute values but also indirectly affected by other behavior events that do not contain them through the neighbor association of attribute values. We divide each behavior into multiple attribute values and then analyze each behavior from the attribute level, where one behavior can be regarded as a set of associations of multiple attribute values in the hidden space. When a pair of attribute values co-occur in multiple behaviors, it means that the two attribute values are more closely relevant.

Event Slice Most previous studies neglected the useful anonymous event interactions from other events that occurred at a relatively recent time of itself. The anonymous event interactions might help obtain contextual semantics in the task of identifying criminal behaviors. Hence, we adopt a chain-based module to learn the interaction patterns between behaviors. We regard (b_i, b_{i+1}, r) as an edge where $b_i \in B$ and r depends on the type of event b_i and b_{i+1}. The edge (b_i, b_{i+1}, r) is directed (i.e., the early event points to the late event). A directed graph allows us to use only the information that has happened in reality. Let $\mathcal{G}_v = (\mathcal{N}_v, \mathcal{E}_v, \mathcal{R}_v)$ be the event slice, $\mathcal{N}_v = \{b_i | b_i \in B\}$ be the set of events, $\mathcal{E}_v = \{(b_i, b_{i+1}, r) | b_i \in B\}$ be the edge set, and $\mathcal{R}_v$ be the relation set.

Agent Slice The agent typically refers to an entity or user carrying out specific behaviors. In cybercrime detection, these agents can be virtual computer programs, specific network ports, or even real user IDs. For instance, in an online payment scenario, users can be considered agents, while a particular port in traffic detection can be designated as an agent. In a broader context, a category of behaviors with a certain characteristic can also be viewed as behaviors initiated by a virtual agent. The selection of agents should enable behavioral analysis of collected data, identifying potential patterns of cybercrime behavior, and comparing them with known criminal behavior patterns to promptly detect and mitigate potential security risks. A major difficulty is that less information among agents is provided to reflect the relations among agents in most scenarios. Thus, we design a flexible module to learn relations between agents in the form of a graph structure.

Given a behavior b, we define a mapping function $\Lambda(\cdot)$ to get the corresponding agent g of b. We denote the node corresponding to g in graph $\mathscr{G}_g$ as n_g. The agent information can be represented as a set of candidate neighbors $\mathscr{C}_g$ for each agent g. To select the appropriate relations between one agent and its candidates, we first compute the similarity between the initial representation $\vec{n}_{g_i} = \Phi[n_{g_i}]$ of agent g_i and its neighbors' initial representation $\vec{n}_{g_j} = \Phi[n_{g_j}]$ as follows:

$$e_{g_i,g_j} = \frac{\vec{n}_{g_i} \cdot \vec{n}_{g_j}}{||\vec{n}_{g_i}|| \cdot ||\vec{n}_{g_j}||} \text{ for } g_j \in \mathscr{C}_{g_i}. \tag{6.1}$$

Then, we select the relation corresponding to the top k such similarities to generate the adjacency matrix A as follows:

$$A_{g_i,g_j} = 1\,\{j \in \text{Top}K(\{e_{g_i,g_k} | k \in \mathscr{C}_{g_i}\})\}, \tag{6.2}$$

where Top$K(\cdot)$ returns the indices of top k values among its input. The value of k can be chosen by the user according to the desired sparsity level.

We denote the agent slice as the graph $\mathscr{G}_g = (\mathscr{N}_g, \mathscr{E}_g, \mathscr{R}_g)$, where $\mathscr{N}_g = \{g = \Lambda(b_i) | b_i \in B\}$ is the node set containing all agents, $\mathscr{E}_g = \{(g_i, g_j, r) | g_i \in \mathscr{N}_g\}$ is the edge set if $A_{g_i,g_j} = 1$, and $\mathscr{R}_g$ is the set of relations between two agents.

Representation Fusion on Each Slice Assume that we have three types of initial representations as input: $\vec{n}'_t = \Phi[n'_t]$, $\vec{n}'_v = \Phi[n'_v]$, and $\vec{n}'_g = \Phi[n'_g]$, where $\vec{n}'_t$, $\vec{n}'_v$, and $\vec{n}'_g$ belong to the embedding space $\mathbb{R}^{d'}$. Then, we utilize the graph-level representation learning module $GNL(\cdot)$ to get the latent representations, as shown below:

$$\vec{n}_i = GNL_{\theta_i}(\vec{n}'_i, \mathscr{G}_i) \text{ for } i \in \{t, v, g\}, \tag{6.3}$$

where GNL_θ is known as the module with θ being the parameters, and the learned latent representations are $\vec{n}_t$, $\vec{n}_v$, and $\vec{n}_g$, respectively. The representation $\vec{n} \in \mathbb{R}^d$ is obtained by $GNL_\theta : \mathbb{R}^{d'} \mapsto \mathbb{R}^d$, where all leaned representations share the same dimensionality d and $\vec{n}'$ is regarded as the input embeddings in GNL_θ.

6.4.3 Sliced Representation Fusion

We present three different extractors (as shown in Fig. 6.5) to learn the behavioral pattern of each agent in a slice-by-slice structure from attributes to events to agents.

Event-attribute extractor enhances the event representations by aggregating information from the representations of attributes. A common means is to use a

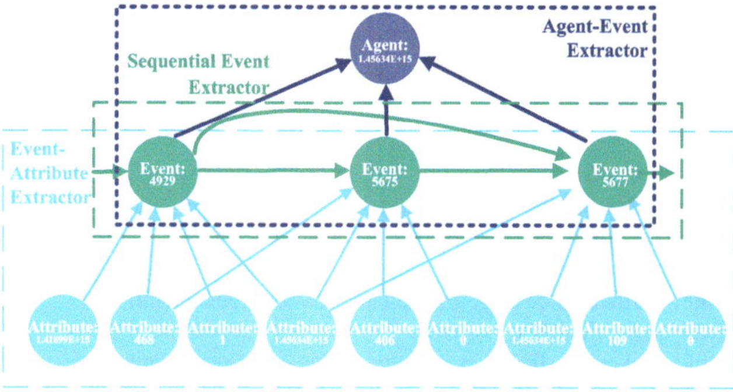

Fig. 6.5 Illustrations of three extractors: event-attribute extractor, sequential event extractor, and agent-event extractor

dense network or concatenate the attribute representations. However, a key issue is that the above methods could not effectively extract the attribute information of each behavior because they are incapable of improving the event representations with interassociations over all behaviors. Inspired by the success of the graph neural network [45], we adopt a graph-based event-attribute extractor that aggregates both the interassociations and intra-associations from attributes to events. It is formulated as

$$\vec{n}_v := GNL_{\theta_{v,t}}((\vec{n}_v, \vec{n}_t), \mathscr{G}_{v,t}). \tag{6.4}$$

On the right side of Eq. (6.4), the representation $\vec{n}_v$ has learned the interassociations and intra-associations during graph-level representation learning on $\mathscr{G}_t$, and the learned information is aggregated in event representations by implementing $GRL(\cdot)$ with parameter $\theta_{v,t}$ on $\mathscr{G}_{v,t}$. There are two kinds of nodes: events and attributes, in which all edges point to events from attributes. Therefore, $\mathscr{G}_{v,t}$ can be formalized as $(\mathscr{N}_{v,t}, \mathscr{E}_{v,t})$, where $\mathscr{N}_{v,t} = \mathscr{N}_v \cup \mathscr{N}_t$, $\mathscr{E}_{v,t} = \{(b_i^j, b_i) | b_i \in B, b_i^j \in b_i\}$. The one-way edge from attributes to events ensures that only attribute information will be aggregated into the event representations.

Sequential event extractor mines sequence information from the agent's historical events. Many methods adopted the chain-based mechanism as the sequential event extractor, which emphasizes the impact of the most recent event on the current event. To capture the pair-wise interactions, we define a time window size w and build a sequential event slice for all events. Given an agent's event sequence $B_g \subset B$, let $\mathscr{G}_{s,v} = (\mathscr{N}_{s,v}, \mathscr{E}_{s,v})$ be the sequential event slice, where $\mathscr{N}_{s,v} \subset \mathscr{N}_v$ is the node set of events in B_g, and $\mathscr{E}_{s,v} = \{(b_{i-k}, b_{i-j}) | b_i \in B_g\}$ denotes the edge set, in which $0 \leq j \leq k \leq w$. The event-attribute extractor is formulated as

$$\vec{n}_v := GNL_{\theta_{s,v}}(\vec{n}_v, \mathscr{G}_{s,v}). \tag{6.5}$$

Algorithm 4: BT: detecting malicious behaviors

Input: The set of events B, the mapping function $\Lambda(\cdot)$ from events to agents
Output: The set of scores S

Obtain the initial representations $\vec{n}_t'$, $\vec{n}_v'$, $\vec{n}_g'$ based on lookup embedding;
foreach $i \in \{t, v, g\}$ **do**
  Build $\mathcal{G}_i$ based on B and $\Lambda(\cdot)$;
  Learn the representation $\vec{n}_i$ based on $GRL_{\theta_i}(\vec{n}_i', \mathcal{G}_i)$;
end
Build the event-attribute slice $\mathcal{G}_{v,t}$, and update the event representations $\vec{n}_v := GRL_{\theta_{v,t}}((\vec{n}_v, \vec{n}_t), \mathcal{G}_{v,t})$;
Build the sequential event slice $\mathcal{G}_{s,v}$, and update the event representations $\vec{n}_v := GRL_{\theta_{s,v}}(\vec{n}_v, \mathcal{G}_{s,v})$;
Build the agent-event slice $\mathcal{G}_{g,v}$, and update the agent representations $\vec{n}_g := GRL_{\theta_{g,v}}((\vec{n}_g, \vec{n}_v), \mathcal{G}_{g,v})$;
Optimize the model;
Initialize $S = \emptyset$;
foreach $b_i \in B$ **do**
  Compute the score s_i according to Eq. (6.7);
  Add s_i into S;
end
Return scores S.

By setting a small w, we can avoid too many long-term historical interactions (e.g., the irrelevant behaviors are helpless or noise) during aggregating the representations.

Agent-event extractor obtains the final representation of each agent by integrating all the representations of its behavioral events. A feasible approach is to model the agent's representations by mean pooling, which essentially treats each event generated by the agent as equally important. The contribution of each event varies when it comes to detecting behaviors. For instance, the agents may pay more attention to a behavior that has been detected by the agents with similar preferences. Motivated by the attention-based mechanism, we learn the agent representations by introducing the learned agent representations.

For each $b \in B$, we can get its corresponding agent $g = \Lambda(b)$. We let $\mathcal{G}_{g,v} = (\mathcal{N}_{g,v}, \mathcal{E}_{g,v})$ be the agent-event slice, where $\mathcal{N}_{g,v} = \mathcal{N}_g \cup \mathcal{N}_v$ and $\mathcal{E}_{g,v} = \{(b, \Lambda(b)) | b \in B\}$. The agent-event extractor is formulated as

$$\vec{n}_g := GNL_{\theta_{g,v}}((\vec{n}_g, \vec{n}_v), \mathcal{G}_{g,v}). \tag{6.6}$$

On the right side of Eq. (6.6), the representation $\vec{n}_g$ has learned the representations of agents with similar preferences, and the representation $\vec{n}_v$ has learned the sequential behavior patterns of agents. With the agent and event representations, our agent-event extractor models the sequence dependence by reducing the noise in the sequence of agents' all behavior.

6.4.4 Malicious Behavior Detection

To identify whether a behavior is cybercrime or not, we incorporate the correlation between learned agent representations and event representations. We concatenate agent representations $\vec{n}_g$ and event representations $\vec{n}_v$ as $[\vec{n}_g, \vec{n}_v]$. Then, we feed $[\vec{n}_g, \vec{n}_v]$ into the artificial neural network (ANN), such as multilayer perceptron, to get the output judgment $\widehat{y}$:

$$\widehat{y} = \text{Sigmoid}(\text{ANN}([\vec{n}_g, \vec{n}_v])). \tag{6.7}$$

We regard the task of identifying criminal behaviors as the binary classification problem of identifying criminal events that camouflage themselves and evade inspection. We adopt the cross-entropy loss function and minimize the negative log-likelihood:

$$Loss = -\frac{1}{|B|} \sum\nolimits_{(b,y)\in B} (y \log \widehat{y} + (1-y) \log (1-\widehat{y})),$$

where y is the true label of behavior b. We utilize Adam [52] to adaptively adjust the learning rate during the training process. Algorithm 4 presents the procedure of BT.

6.4.5 Graph-Level Representation Learning

Here, we present how to encode the information in graph. Let $\mathbb{R}^d$ be an intermediate representation space, and the graph level representation learning module aims to obtain representations $\mathrm{H} \in \mathbb{R}^d$. Our $GRL(\cdot)$ is based on graph neural networks due to their success in real applications [40, 49]. We stack multiple neural network layers to build $GRL(\cdot)$ for getting the node representations of the whole graph. Here, we denote the output of the lth neural network layer as H^l and describe the lth layer that consists of two components: information propagation and information aggregation.

Information Propagation Nodes' information is affected by their local neighborhood information, so the local neighborhood information may effectively help improve the subsequent tasks. For a node, the mean pooling operation can play as a straightforward solution to obtain the local neighborhood information. We consider utilizing an attention mechanism to distinguish the importance of different neighborhood information since not all of the information is conducive to learning the representations. Thus, we denote the basic operator $Attention$ to estimate the importance of each node in its neighbor node set and $Message$ to pass information from source nodes to target nodes.

In this chapter, we adopt the heterogeneous mutual attention mechanism in heterogeneous graph transformer (HGT) [53]. It is worth noting that different attention mechanisms can be used here. To fully utilize the heterogeneous properties in graphs, e.g., multiple node type and edge type, we adopt HGT to act as the information propagation module. Formally, HGT calculates the multi-head attention for each edge (s, t, r) that consists of source node s, target node t, and relation r, by:

$$\begin{aligned} Attention(s,t,r) &= \underset{\forall s \in N(t)}{Softmax}\left(\underset{i \in [1,h]}{\Vert} Att^i(s,t,r) \right), \\ Att^i(s,t,r) &= \left(K^i(s) W_r^A Q^i(t)^T \cdot \frac{\mu_r}{\sqrt{d}} \right), \\ K^i(s) &= \text{K-Linear}_r^i \left(\mathrm{H}^{l-1}[s] \right), \\ Q^i(t) &= \text{Q-Linear}_r^i \left(\mathrm{H}^{l-1}[t] \right), \end{aligned} \tag{6.8}$$

where W_r^A is a matrix for each relation type r, μ_r is a prior tensor to denote the general significance of relation type r, and Att^i is the ith attention head. HGT projects the source node s (target node t) with a linear projection K-Linear (Q-Linear) into the ith key (query) vector. Then, HGT calculates the similarity between the query vector $Q^i(t)$ and key vector $K^i(t)$.

For each edge (s, t, r), HGT calculates its multi-head $Message$ by

$$\begin{aligned} Message(s,t,r) &= \underset{i \in [1,h]}{\Vert} Msg^i(s,t,r), \\ Msg^i(s,t,r) &= \text{M-Linear}_r^i \left(\mathrm{H}^{l-1}[s] \right) W_r^M, \end{aligned} \tag{6.9}$$

where W_r^M is a matrix for incorporating the relation dependency, and the linear projection M-Linear$_r^i$ projects the type of relation into the ith message vector in all message heads.

Information Aggregation We aggregate the representations of neighborhoods and itself by the attention-based importance, which is denoted as $Aggregate$. In this chapter, inspired by Hu et al. [53], we implement $Aggregate$ as follows,

$$\begin{aligned} Aggregate(s,t,r) &= \text{A-Linear} \left(\sigma (\widetilde{\mathrm{H}}^l[t]) \right) + \mathrm{H}^{l-1}[t], \\ \widetilde{\mathrm{H}}^l[t] &= \underset{\forall s \in N(t)}{\oplus} (Attention(s,t,r) \cdot Message(s,t,r)), \end{aligned} \tag{6.10}$$

where $\sigma(\cdot)$ is a nonlinear activation function, and A-Linear is a linear projection.

Finally, we can get the Lth layers' output representation $\mathrm{H}^L[t]$ for the target node t by stacking the L layers. The feature representation learner can be denoted as follows:

$$GRL(\cdot) = \mathrm{H}^L[t] \leftarrow Aggregate\,(Attention(\cdot), Message(\cdot))\,.$$

6.5 Evaluation of BT

We employ the cybersecurity event datasets (fraudulent transaction dataset on online payment services and malicious traffic dataset on anonymous network communications) in representative cybercrime scenarios and evaluate five aspects. We provide a roadmap of this section:

- **Performance comparison.** We compare the performance of BT and existing representative works in identifying criminal behaviors (in Sect. 6.5.2).
- **Ablation study.** We illustrate the gain brought by our customized components of BT in identifying criminal behaviors (in Sect. 6.5.3).
- **Visualization study.** We visualize the distribution of the learned representations (including attribute representations, event representations, and agent representations) (in Sect. 6.5.4).
- **Case study.** We exhibit the consistency of behaviors and learned representations in representative cases (in Sect. 6.5.5).
- **Parameter study.** We evaluate the sensitivity of our method over key parameters (in Sect. 6.5.6).

6.5.1 Experimental Setup

6.5.1.1 Datasets and Implementation

We evaluate our method over the representative datasets (OnlinePayment and CICDarknet2020) in two typical applications, i.e., fraudulent transaction detection in online payment services and suspicious traffic detection in virtual private networks. These datasets contain representative cybercrimes, i.e., fraudulent online payment behaviors in OnlinePayment[2] and malicious network traffic behaviors in

[2] **OnlinePayment**: Privacy is an essential concern in anti-fraud scenarios. Due to the constraints of privacy protection and business confidentiality, the dataset used in this study is not publicly available. To ensure the privacy of individuals and safeguard business secrets, the data has undergone desensitization before being used for research purposes. The distribution of the dataset is regulated by the bank, and researchers are required to adhere to the bank's Developer Agreement and Policy prior to accessing the dataset.

CICDarknet2020[3] [54]. The information in these datasets provides the various features of behavior, which can be regarded as the phenomenological description (e.g., quantitative characterization) of behaviors [55]. For instance, OnlinePayment provides the complete users' operation records during the payment process, and CICDarknet2020 collects a set of important network flow features to trace behaviors of both users and attackers that generate malicious traffic by interacting with systems.

OnlinePayment dataset is a real-world online banking payment transaction dataset from one of the biggest commercial banks in China. Online payment service providers face serious threats from malicious criminal behaviors. The existing methods have made considerable progress in detecting cybercrimes. Unfortunately, attackers adopt meticulously planned strategies to conceal their malicious intent, making it challenging to distinguish their malicious behavior from benign behavior in order to evade detection systems. Some of them employ a deceptive strategy known as "throwing a stone to clear the road" to create a facade of normalcy (i.e., they usually perform several benign actions before revealing their attacking intents) [56]. Some of them incite normal users to participate in negligently set promotional rush buying activities to cover up their behavior [57]. Even worse, these strategies are cheap, easy to use, and allow attackers to circumvent regulations. In this chapter, we sampled approximately 60,000 payment transactions from the raw data. In these transactions, attackers primarily employed tactics such as telephone scams, phishing, and coupon clipping within organized groups to disguise their malicious activities. For each transaction, we mainly recorded a transaction ID and eight fields to construct our model, including account number, merchant number, place number, timestamp, amount, IP verification, final outcome, and transaction type, as summarized in Table 6.2. Within these transactions, all users initiated a variety of transaction activities across 375 different merchants and 316 different locations, with varying amounts involved. In our testing, we did not specifically identify the exact types of transaction activities. Instead, our focus was on labeling transactions as fraudulent or normal. We report the average results obtained from 10 independent runs.

CICDarknet2020 dataset is a traffic dataset generated by the Canadian Institute for Cybersecurity, which detects and characterizes VPN and Tor applications together as the real representative of darknet traffic by amalgamating two public datasets, to create a complete darknet dataset covering Tor and VPN traffic, respectively. The darknet can offer notable hidden services in underground cybercrimes. So many worldwide users are connected via bridge to hidden services (such as Tor), where about 20% of these services are suspicious [54]. Analyzing darknet traffic helps in the detection of malicious illegal activities. For example, illegal trading often tops the hidden services list with the highest number of closely

[3] **CICDarknet2020**: It is a traffic dataset publicly released by the Canadian Institute for Cybersecurity. You can access the dataset through the following link: www.unb.ca/cic/datasets/darknet2020.html.

Table 6.2 Selected features in OnlinePayment dataset

Field	Role	Description
Account ID	Agent	The card number of transaction, the unique identification of account
transaction ID	Event	The unique identification of transaction
Transaction time	Attribute	The time when the transaction happened, beginning with year, accurate to second
Transaction amount	Attribute	The amount of money in the transaction
Transaction type	Attribute	The type of the transaction
Merchant number	Attribute	Each merchant number represents a merchant in a transaction
Place number	Attribute	The issuing place of banking card
Frequent IP	Attribute	The status indicating whether the current IP is commonly used by the account
Check	Attribute	The verification tool at the process of the transaction

related services. The dataset has 83 features and 141,530 rows, where each sample is labeled as benign or malicious and these features are generated by deploying ISCXFlowMeter to read and process PCAP files. The dataset contains one benign class of 117,219 samples and eight different malicious classes of 24,311 samples. The malicious darknet traffic constitutes Audio Stream, Browsing, Chat, Email, P2P, Transfer, Video Stream, and VOIP, which are generated by special applications. In our work, the traffic from illegal hosts on the darknet is treated as malicious criminal behavior. We conduct our method based on nine selected features (as shown in Table 6.3) that are chosen as the most important attributes in a simple tree-based model. In each independent experiment, we deployed our method on a randomly selected 20% of the data and reported the evaluation results after 10 repetitions.

We split all the data into training set and testing set with a ratio of 70 : 30. In the training set, we pick 20% of the samples as the validation set. We report the average results after 10 independent runs on the testing set. The experiments are conducted under Windows 10 Pro operating system with Intel i7-9700K @ 3.6 GHz CPU, 64 GB RAM, and GeForce RTX 2080 Ti Graphics card.

6.5.1.2 Competing Methods

Our method is compared with several popular methods, including linear regression (LR), support vector machine (SVM), multilayer perceptron (MLP), CatBoost (CatB) [58], SAINT [59], Devnet [60], HGT [53], TabNet [36], and BIG [48]. We use the implementation of linear regression and support vector machine available at the scikit-learn Python package. Multilayer perceptron is built upon torch as the backend. CatB, SAINT, DevNet, HGT, TabNet, and BIG come from the implementation on the third-party open-source platform.

During the experimental deployment, the first six methods are based on behavioral events, where we directly feed the model with behavioral events consisting of

Table 6.3 Selected features in CICDarknet2020 dataset

Field	Role	Description
Idle-Max	Agent	The maximum amount of time a network flow was idle before becoming active
traffic ID	Event	The unique identification of traffic sample
Fwd-Seg-Size-Min	Attribute	The minimum forward segmentation size of a data packet
Bwd-Packet-Length-Min	Attribute	The minimum length of a network packet during transmission in the backward (receiving) direction
Protocol	Attribute	The type of transmission protocol used by network packets at the transport layer
Idel-Mean	Attribute	The mean amount of time a network flow was idle before becoming active
FWD-Init-Win-Bytes	Attribute	The initial sliding window size value when the connection is established in the forward direction
FIN-Flag-Count	Attribute	The number of segments with the end flag FIN set in a network flow or connection
Sub-flow-Bwd-Bytes	Attribute	The total amount of traffic size of each subtraffic in a network flow during the receiving process
Bwd-Packet-Length-Max	Attribute	The maximum length of a packet during transmission in the backward (receiving) direction in a network flow or connection

multiple attribute values to obtain corresponding behavioral judgments. Regarding the graph-based model, HGT is an advanced graph neural network method that excels in node classification tasks. Leveraging the original behavioral data, we construct a heterogeneous graph comprising multiple nodes representing behaviors and their associated attributes. For each pair of (behavior, one attribute) present in a behavior, we define it as an edge. The graph used in HGT establishes an intuitive link between behaviors, indicating that behaviors sharing common attributes exhibit similar contexts within the graph structure. BIG is a graph-based behavioral modeling solution that delves into associations across multiple hierarchical levels of behaviors. We retain the attribute slices as the property graph for BIG and establish the required correspondence between events and attributes based on the event-attribute slice.

6.5.1.3 Evaluation Metrics

We use the precision (Prec), true positive rate (TPR), and false positive rate (FPR) over the datasets and their ground labels to evaluate the performance of our method and baseline models. In addition, we use two popular metrics, the area under the receiver operating characteristic curve (AUC-ROC) and the area under the precision-recall curve (AUC-PR), to have a complementary evaluation. AUC-ROC is widely adopted to evaluate binary detection tasks [61], where a large AUC-ROC value indicates the model has better performance. Note that most datasets are unbalanced in detecting cybercrimes. AUC-PR requires excellent performance on the positive class (i.e., malicious criminal behaviors) and does not emphasize the performance on the negative class (i.e., benign one). Compared to AUC-ROC, AUC-PR is less affected by the imbalance of abnormal data. Thus, it is suitable and necessary to adopt AUC-PR as the choice of metrics. Drawing inspiration from [62], we use average precision to calculate AUC-PR. Similar to AUC-ROC, large AUC-PR indicates better performance. Meanwhile, according to the industry requirement, the detection models need to pay more attention to improving the TPR under the low FPR to avoid disturbing normal behaviors. If we solely focus on achieving a high TPR while disregarding the FPR, we may end up wrongly identifying a large number of normal behaviors as malicious, leading to a high rate of false alarms that unnecessarily trouble users and the system. So we also record the performance of models under low FPR in this work. Taking into account the tolerance of the business toward disturbances caused by malicious behaviors, we measure the performance of the OnlinePayment dataset at the FPR ranging from 0.05% to 0.1% and the performance of the CICDarknet2020 dataset at the FPR ranging from 3% to 5%.

6.5.2 Effectiveness in Real-World Datasets

The detection performance of our method and baselines are summarized in Tables 6.4 and 6.5. The performance of BT on the AUC metric surpasses all baseline methods, and it can achieve high TPR at a low FPR level. Regarding the proportion of malicious behaviors, the percentage of malicious behaviors in the OnlinePayment dataset is under one percent, while the percentage of malicious behaviors in the CICDarknet2020 dataset is approximately 20 times higher than that in the OnlinePayment dataset. The higher percentage of malicious behaviors in the CICDarknet2020 dataset is beneficial for establishing malicious behavior patterns using existing approaches, allowing most methods to achieve comparable performance levels. By mining information at the microscopic level, BT is able to build robust behavioral models even with limited exposure to malicious behaviors. As such, BT can achieve high detection capabilities for malicious behaviors, significantly outperforming baseline methods on the OnlinePayment dataset.

When evaluating the performance of different methods on the OnlinePayment dataset, it is found that all methods achieved high AUC-ROC values (above 0.86). This can be attributed to most methods being able to capture most malicious behaviors when allowed more disturbances of normal behaviors. However, the practical significance of these results may be limited. To gain further insights, the performance of the methods is also assessed using AUC-PR. Remarkably, the BT method outperforms all baseline approaches in terms of AUC-PR. The improvement achieved by BT was significantly greater than that of CatB (22%), MLP (53%), and SAINT (104%). This indicates the superiority of the BT method in capturing and distinguishing malicious behaviors for fraud detection. While SAINT, DevNet, and

Table 6.4 Results (AUC) on OnlinePayment and CICDarknet2020

	Baseline	AUC-PR	AUC-ROC
OnlinePayment	LR	0.2890	0.9571
	SVM	0.2600	0.9599
	CatB	0.7490	0.9886
	SAINT	0.4495	0.8602
	DevNet	0.3760	0.8952
	HGT	0.8250	0.9919
	TabNet	0.7018	0.9927
	BIG	0.7401	0.9880
	BT	0.9199	0.9940
CICDarknet2020	LR	0.6744	0.9184
	SVM	0.7075	0.9192
	MLP	0.7411	0.9370
	CatB	0.8246	0.9558
	SAINT	0.6731	0.9054
	DevNet	0.4258	0.7716
	HGT	0.8679	0.9658
	TabNet	0.8180	0.9626
	BIG	0.8890	0.9777
	BT	0.8938	0.9881

HGT have lower AUC values than MLP and CatB, they have better performance at lower FPR, which is important in practice. They achieve performance gains by utilizing more information (such as the distribution of transaction behaviors). The results show that an extremely imbalanced proportion of malicious behaviors will make the identification of malicious behaviors difficult, especially at low FPR. Notably, even when other competing methods exhibited poor performance with an FPR of 0.05%, our proposed method is able to successfully identify 85% of positive samples (i.e., malicious behaviors). This noteworthy result further emphasizes the advantages of our work in mining association information at the microscopic level and capturing complex dependencies between behaviors.

In the context of identifying untrustworthy network traffic on the CICDarknet2020 dataset, the proportion of malicious behaviors is relatively high, accounting for 17.09% of the overall dataset, so even traditional classification methods can achieve AUC-ROC values exceeding 0.9. However, we also observe that the TPR of these methods is limited at low interference rates, approximately 65%, meaning general models tend to pursue overall performance on all samples by achieving accurate classification at high FPR, neglecting model performance at low FPR. Most baseline methods focus on associations between behavior and labels at the macrolevel, performing poorly at low FPR. Furthermore, we can observe that some baselines mining richer information from the data (e.g., graph-based methods exploring correlations between samples) achieve not only better AUC but also capture over 80% of malicious behaviors at low FPR. This confirms that latent

Table 6.5 Results under different FPRs on OnlinePayment and CICDarknet2020

		FPR=0.05%			FPR=0.75%			FPR=0.10%		
	Baseline	TPR	Prec	F1	TPR	Prec	F1	TPR	Prec	F1
OnlinePayment	LR	8.33%	52.63%	0.1439	15.00%	58.06%	0.2384	15.83%	31.15%	0.2099
	SVM	11.67%	58.33%	0.1944	12.51%	51.72%	0.2015	29.16%	66.03%	0.4045
	MLP	8.33%	43.48%	0.1399	9.16%	45.87%	0.1527	24.16%	63.04%	0.3493
	CatB	6.67%	42.11%	0.1151	47.51%	81.43%	0.6001	67.51%	81.82%	0.7398
	SAINT	37.24%	87.10%	0.5217	44.82%	82.28%	0.5803	46.21%	77.90%	0.5801
	DevNet	12.00%	62.07%	0.2011	18.67%	68.29%	0.2932	22.67%	65.38%	0.3367
	HGT	45.83%	91.37%	0.6104	50.83%	90.22%	0.6502	51.67%	87.15%	0.6488
	TabNet	14.17%	65.39%	0.2329	25.83%	68.89%	0.3757	65.83%	81.44%	0.7281
	BIG	16.67%	71.43%	0.2703	71.20%	80.22%	0.7544	75.46%	86.41%	0.8056
	BT	85.00%	91.43%	0.8810	90.83%	88.61%	0.8971	92.50%	86.05%	0.8916
		FPR=3%			FPR=4%			FPR=5%		
	Baseline	TPR	Prec	F1	TPR	Prec	F1	TPR	Prec	F1
CICDarknet2020	LR	44.61%	75.94%	0.5620	55.72%	74.75%	0.6385	61.58%	70.57%	0.6577
	SVM	48.23%	76.65%	0.5921	58.76%	74.97%	0.6583	69.37%	73.88%	0.7155
	MLP	51.27%	78.51%	0.6203	56.88%	75.18%	0.6476	67.99%	73.89%	0.7081
	CatB	66.94%	81.93%	0.7368	72.23%	79.20%	0.7555	77.14%	76.52%	0.7683
	SAINT	66.15%	75.09%	0.7034	78.76%	77.62%	0.7819	80.32%	74.91%	0.7752
	DevNet	17.39%	53.63%	0.2626	19.43%	50.13%	0.2801	22.39%	48.13%	0.3057
	HGT	69.58%	82.72%	0.7558	75.08%	79.62%	0.7728	82.49%	77.59%	0.7996
	TabNet	66.71%	80.08%	0.7279	69.85%	82.98%	0.7585	81.53%	77.38%	0.7940
	BIG	68.56%	82.72%	0.7498	75.56%	79.78%	0.7761	84.26%	79.79%	0.8196
	BT	69.51%	82.91%	0.7562	78.14%	80.32%	0.7922	84.45%	77.99%	0.8110

information in the data can further help build accurate behavior models. With existing methods already deeply mining data information, our method achieves better performance by leveraging information at the microscopic level. These hard-won performance gains stem from BT's ability to more precisely identify subtle malicious behaviors that are meticulously planned and difficult to recognize.

6.5.3 *Ablation Study*

6.5.3.1 Ablation Setting

We conduct experiments with several variants of BT to further demonstrate the effectiveness of our designs. We examine the importance of the key components of BT by comparing it (denoted as V_0) to its variants (denoted as V_i, i denotes the ith variant). Recall that the default version (V_0) builds a sequential event slice in the sequential event extractor, concatenates the agent representations and the event representations for multilayer perceptron, and has four hidden layers with cross-entropy loss in the prediction.

The first variants are to change the concatenation of agent representations and event representations. Concatenation+ (V_1) further concatenates all attribute representations corresponding to the event. Concatenation- (V_2) only uses event representations as the input of multilayer perceptron. The second variants are to change the structure of the sequential event slice. Compared with Ours, Sequence- (V_3) removes the sequential event extractor and directly applies the agent-event extractor after the event-attribute extractor. Chain (V_4) indicates that each event only points to its most recent subsequent event in the structure of the sequential event slice. Short term (V_5) indicates that all events only point to the last event in a window during building a sequential event slice. The third variants are to change the unit of the prediction. We first remove the nonlinear learning hidden layers in neural network, making it equivalent to learning a direct linear mapping from the original data space to the anomaly score space (linear, V_6). Then, we deployed the multilayer perceptron with two hidden layers (2-Hidden, V_7) and three hidden layers (3-Hidden, V_8). The fourth variants are to change the loss function of the optimization module. We replace cross-entropy loss with focal loss (V_9) [63] or TPAUC loss (V_{10}) [64].

6.5.3.2 Finding

Table 6.6 shows the ablation results on OnlinePayment. We observe that our default model (Ours) significantly outperforms Concatenation+ in AUC-PR, indicating better performance at lower FPR. While Concatenation+ introduces attribute representations, performance does not notably improve and sometimes decreases. This is likely because the high-dimensional concatenated representation misguides the

Table 6.6 Performance of BT and its variants on the OnlinePayment

Ablation	AUC-PR	AUC-ROC
V_0 : Ours	0.9199 ± 0.0433	0.9940 ± 0.0002
V_1 : Concatenation+	0.7818 ± 0.0645	0.9950 ± 0.0012
V_2 : Concatenation-	0.9109 ± 0.0096	0.9903 ± 0.0012
V_3 : Sequence-	0.8159 ± 0.0591	0.9868 ± 0.0004
V_4 : Chain	0.8234 ± 0.1474	0.9863 ± 0.0010
V_5 : Short-term	0.8386 ± 0.1109	0.9903 ± 0.0068
V_6 : Linear	0.8889 ± 0.0025	0.9886 ± 0.0012
V_7 : 2-Hidden	0.8956 ± 0.0868	0.9914 ± 0.0035
V_8 : 3-Hidden	0.8871 ± 0.0456	0.9852 ± 0.0027
V_9 : Focal	0.9549 ± 0.0092	0.9874 ± 0.0033
V_{10} : TPAUC	0.9041 ± 0.0700	0.9890 ± 0.0014

model in event representation due to treating each attribute equally. The results show that blindly feeding attribute representations downstream does not necessarily yield gains. This also demonstrates the effectiveness of our designed event-attribute extractor in incorporating attribute representations into the event representation. The improvement over Concatenation- is limited for Ours. Agent representations with agent preferences contribute to identifying malicious behaviors. However, since event representations already contain most information from behavioral data, they can determine behavioral suspicion through attribute consistency and associations between preceding and following events. Therefore, the gains in malicious behavior detection brought by agent representations are not significant.

The event-event information may help predict the next related events based on a few events in chronological order. In most cases, it improves the detection ability of the model, where the average performance from high to low is: ours, short term (removing the sequential event extractor), chain (each event only impacts its most recent subsequent event), and sequence- (all events impact the last event in a window). The short term version, which removes the sequential event extractor and focuses on capturing the relevance of the last event to each event in a window, ranks second in terms of performance. However, it is worth noting that the chain does not consistently yield beneficial results. Directly utilizing the chain-based event-event information may introduce noise when some of the encoded information is not relevant to the last event. Therefore, the effectiveness of the chain-based version is highly dependent on the specific prediction task. Furthermore, we observe that the model has good performance after refining the data with microscopic co-occurrence relations in most cases. The results show that our method obtains expressive representations by mining the hidden and intricate relations among data, which effectively contributes to a better ability for behavior modeling.

Regarding the choice of loss function, our model achieves similar performance in terms of AUC-ROC and AUC-PR compared to the focal and TPAUC loss functions. Interestingly, focal shows a decent improvement in AUC-PR, which can be attributed to its tailored handling of imbalanced data. The results suggest there is potential for further optimization by selecting a suitable loss function. Our default loss function performs comparably to the specialized alternatives, but incorporating a specifically designed loss function like focal could further enhance performance, especially in producing well-calibrated probability outputs for the minority class. It demonstrates our model's adaptability to more challenging security tasks by addressing data imbalance issues.

6.5.4 Visualization Study

To intuitively reflect the effectiveness of our BT, we visualize the learned representations (including agent, event, and attribute representation) with the help of tSNE [65] on OnlinePayment dataset. The visualization results are shown in Fig. 6.6. As we can see, most attributes of the same type are divided into the same cluster in Fig. 6.6a. This confirms that the basic information of attributes is effectively retained after being transferred on the attribute slice, and there are significant differences between different attributes. On the other hand, the point distribution (in Fig. 6.6a) shows that we learn the relative relationship between any attribute value and other attribute values in the representation space. So an event consisting of a set of attribute values can be viewed as a simple complex composed of several points in the topological space where the shape of the complex can characterize the event. With the help of the event-attribute extractor, we abstract an event into an event representation by aggregating information from the learned representations of attributes. In Fig. 6.6b, we observe that BT is able to separate the representations of events with different labels. This is also the basis for our method to achieve good detection ability. But we also observe that the simple event representation only realizes the binary classification, and it is not outstanding in expressing the behavior, e.g., different anomalies. In our work, each agent is abstracted into an agent representation. In Fig. 6.6c, our method obviously realizes personalized modeling for different agents. Considering that events with the same attributes may represent different behaviors under different agents, we further concatenate the agent representation and event representation, where the visualization results are shown in Fig. 6.6d. Compared with the results in Fig. 6.6b, we can see that our approach further delineates these points, while maintaining the classification abilities for different event labels. Specifically, malicious behaviors launched by different fraudsters are partitioned into distinct fraud clusters. By partitioning fraud events into multiple distinct clusters, this facilitates identifying specific malicious event types (although performance gains may not be significant for simple binary malicious/benign detection). Similarly, more clusters are observed for normal events, corresponding to different behavioral patterns. Overall, agent representation aids understanding complex security problems beyond the scope of two-class detection.

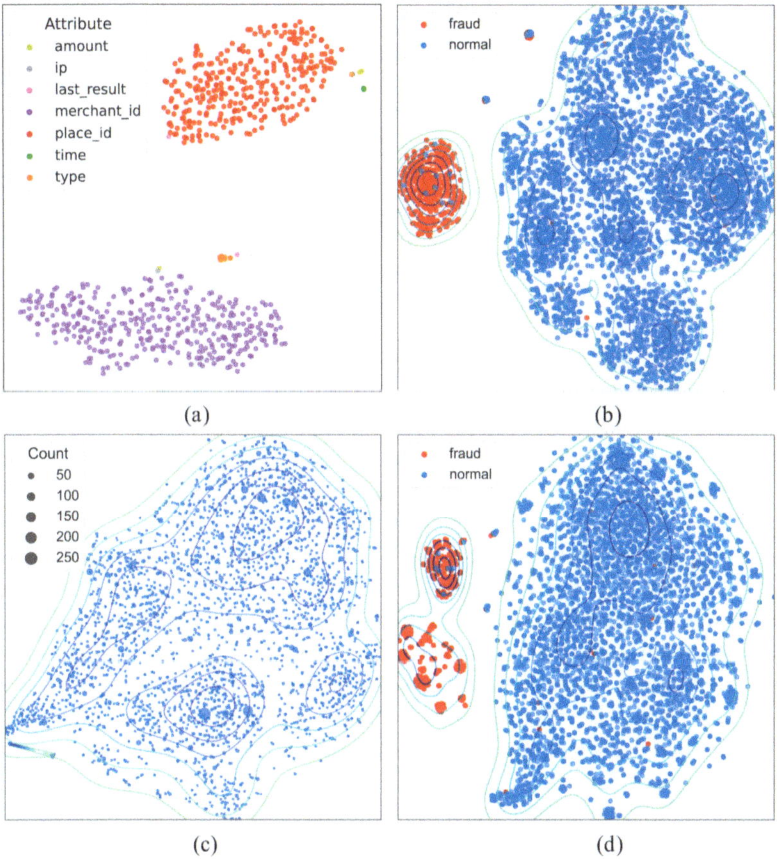

Fig. 6.6 Visualization of the learned representation distribution on the OnlinePayment dataset. (**a**) Attribute. (**b**) Event. (**c**) Agent. (**d**) Agent||Event

6.5.5 *Case Study*

In order to gain a deeper understanding of the learned representations, we select behavioral data from agents in the OnlinePayment dataset to demonstrate the consistency between behavior and representations. Table 6.7 presents the statistical information of the selected agents. We count the most common attribute I and its occurrence frequency J of the agent, denoted as I (J). Figure 6.7 illustrates the similarity between different agents and the similarity between the behavioral events they initiate. The cosine similarity metric is employed to quantify the similarity between event or agent representations. The vertical axis and horizontal axis correspond to the events performed by the respective agents. The events on the horizontal axis belong to one agent, and the events on the vertical axis belong to another agent. The similarity of the representation between two events is visualized

Table 6.7 Statistical information of the selected agents in OnlinePayment

Account ID	The number of behavioral events	Transaction time	Transaction amount	Transaction type	Place number
① 62122*842375	11	1101 (7)	11 (9)	100 (11)	14301 (11)
② 62220*767005	15	1101 (7)	12 (15)	105 (9)	13006 (15)
③ 62220*714378	33	1100 (33)	12 (20)	100 (33)	14402 (33)
④ 62122*941929	27	1110 (27)	12 (23)	100 (27)	13202 (27)
⑤ 62220*688951	13	1101 (13)	12 (12)	100 (13)	12610 (13)
⑥ 62122*335936	19	1110 (9)	13 (19)	100 (19)	10403 (19)

using a color gradient. The similarity value mentioned in the top title denotes the cosine similarity between the two agent representations. The heavier the color is, the higher the similarity between the two event representations. Each row or column corresponds to the similarities between the event representations of one agent and that of another agent. In Fig. 6.7a and b, there is indeed significant event similarity between similar agents. In particular, we observe that the event similarities are generally high. This proves that our model can identify agents with similar events. Furthermore, we selected two highly dissimilar agents (their similarity is about -0.81), where one agent generates multiple abnormal events and the other generates multiple normal events. From Fig. 6.7c, we observe that their corresponding events are different, that is, most of them show negative similarities. Considering that the behaviors generated by different agents with similar events may not be similar, we further explored the behaviors between unrelated agents (their similarity is about 0), where the result is shown in Fig. 6.7d. In raw data, the two agents' frequently visited merchant_id, payment amounts, and payment periods are obviously different, which is consistent with the fact that the agent similarity is close to 0. From Fig. 6.7d, we see that the events between the two agents are obviously dissimilar in three parts. Agent1 (corresponding to the vertical axis) has spent 19 times in three different periods in the same merchant. Therefore, we can see that each event of Agent2 (corresponding to the horizontal axis) is very similar to the above 19 events. Unlike Agent1, Agent2 tends to have fixed payment behaviors with one frequently visited merchant and three occasionally visited merchants. This phenomenon is consistent with several parts divided by three dissimilar events in Fig. 6.7d. To sum up, these cases suggest that our model is able to learn the latent behavior by learning the representations.

6.5.6 *Parameter Study*

Our method involves a number of parameters that may affect the detection performance. In order to prove that our method can be applicable to different changes of parameters, we use six values of each key parameter on the malicious behaviors detection. Specifically, we vary the size of time window (w) in the sequential

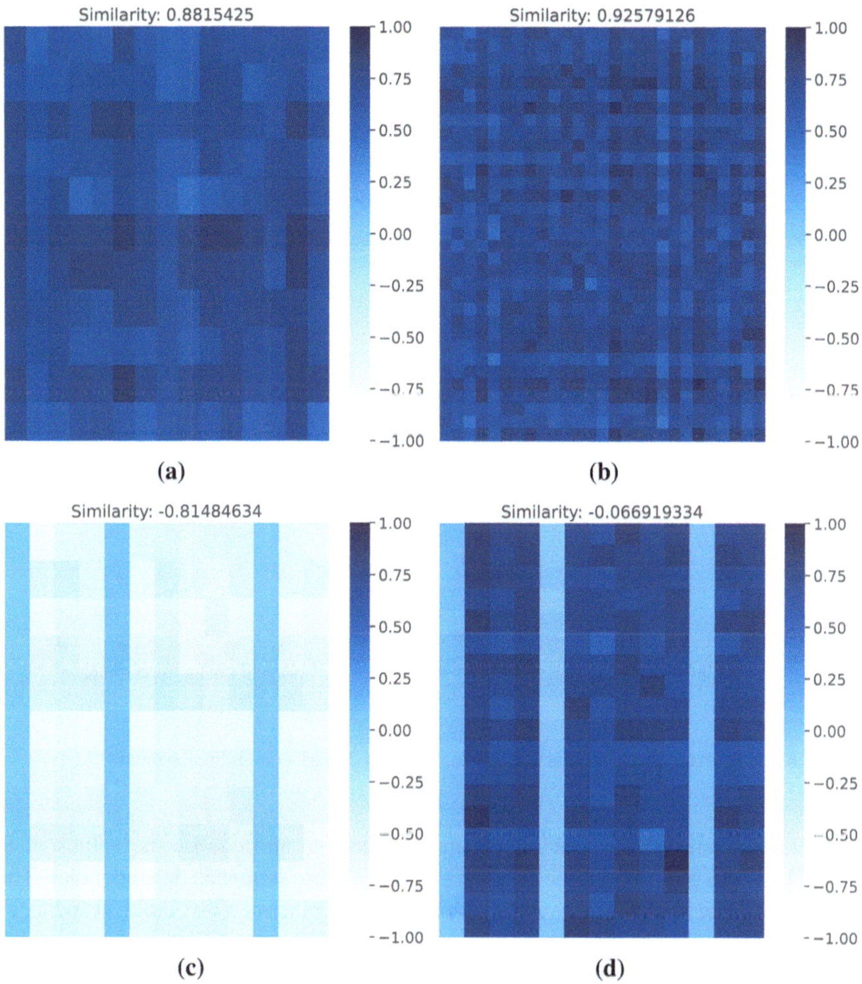

Fig. 6.7 Illustration of the similarity of event representations between different agents. (**a**) Similar normal events by similar agents (① and ②). (**b**) Similar abnormal events by similar agents (③ and ④). (**c**) Malicious and benign events by dissimilar agents (① and ⑤). (**d**) Partially similar events by unrelated agents (⑤ and ⑥)

event extractor, the number of attention heads (h) in the graph-level representation learning, the dimension of input representation (d') in the lookup embedding, and the dimension of output representation (d) in the intra-slice representation fusion, to determine their impacts on malicious behavior detection. Except for the key parameters being tested, all other parameters assume default values.

We show the parameters study on OnlinePayment in Fig. 6.8. We observe that the performance is impacted by the size of the time window (w), but the improvements are not linear with w. Our method can learn sequence dependencies to improve the

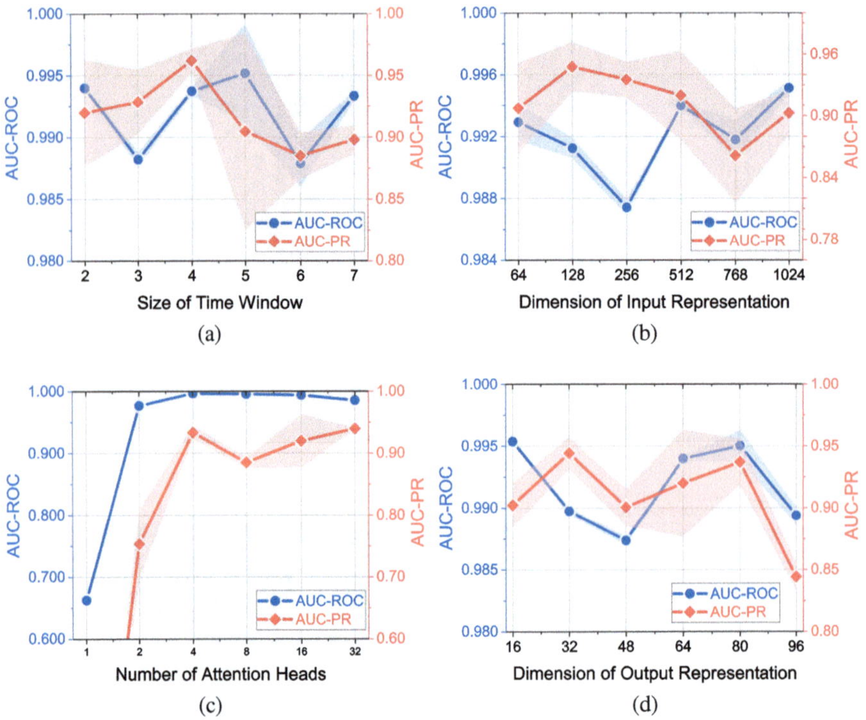

Fig. 6.8 Performance w.r.t. the parameters on $4 \times 6 = 24$ tasks on OnlinePayment

model. We examine the size of the time window w and choose the values of w from 2 to 7. In general, a larger w corresponds to more sequence dependencies. The performance increases as w and then drops after w reaches a threshold. It indicates that our method is able to learn more sequence dependencies to improve the model. The noise in sequence dependencies may weaken the performance when w is too high. So the performance increases as w and then drops after w reaches a certain threshold. It is because, when w is small, information from sequence dependencies may be partially captured to establish an accurate agent model, while when w is too high, the noise in sequence dependencies will weak the performance.

The results about multi-head attention depict a first-increasing and then-steady line, which implies that appropriately increasing the number of attention heads can learn richer dependencies to improve performance. Multi-head attention allows the graph-level representation learning layer to learn relevant information in different representation subspaces. Therefore, appropriately increasing the number of heads can learn richer dependencies within the input graph, help to capture hidden information in the graph structure to improve performance for detecting deviations. For OnlinePayment dataset, the AUC values initially improve with the number of attention heads (h) but further improvements are slower beyond a certain threshold ($h \geq 4$).

For the dimension of input/output representations, we observe that the model relatively performs weakly when the dimension of input/output representations (d' or d) goes beyond 512 or 80. This is because the large d' leads to sparse initial embeddings, which aggravates the learning burden. Similar to d', the appropriate dimension of output representation can provide adequate information for multilayer perceptron, which is conducive to identifying malicious behaviors in most cases. When the dimensions d' and d are small, the training speed of our model will be faster than that with a large dimension. This does not mean that our method can obtain better representations in a fast way, since the low-dimensional representation cannot fully represent complex dependencies, resulting in the loss of information. For OnlinePayment dataset, our model achieves acceptable performance when the dimension is small. Taking both prediction performance and computational time into consideration, we do not set d' or d in a large value.

6.6 Discussion

6.6.1 Security Against Adversarial Attacks

Adversarial attacks pose a significant threat to learning-based security models. Attackers can deceive models by making subtle modifications to their own behavioral data, thereby manipulating data during training and inference stages to induce incorrect predictions. To address these issues, when constructing behavior slices, we mitigate adversarial attacks to some extent by binning data and learning its semantics. Firstly, by grouping the original data into different intervals or categories, we can reduce the impact of adversarial attacks. Through binning, the model relies more on the data distribution characteristics within each bin rather than individual data points, thereby decreasing the likelihood of attackers deceiving the model through minor modifications. Secondly, by deeply understanding the semantics and contextual information of the data rather than directly computing the raw values, the model can better capture the relationships between data, enhancing its robustness against adversarial attacks. Learning the semantics of the data helps the model better understand the true meaning of the data, reducing the potential for misleading effects brought about by adversarial attacks. Additionally, our graph-level learning module utilizes a generic GNN model, allowing us to further enhance the model's resistance to adversarial attacks by combining existing model robustness methods such as adversarial training, defensive training, input reconstruction, and others. This comprehensive strategy helps strengthen the model's robustness and effectively address the evolving threats posed by adversarial attacks.

6.6.2 Real-World Deployment

Although BT can accurately detect cybercrimes, it is not comprehensive enough in practical deployment. Unlike academic models, real-world models tend to adopt a multilayered architecture to achieve comprehensive security protection from development environments to operational business systems. For example, BT relies on structured behavioral data, which needs to be extracted and cleaned from unstructured records such as business logs. In addition, in the identification of cybercrimes, there is limited tolerance for missed detections, and a higher recall rate is pursued. Therefore, our model is more suitable as a second line of defense, working in conjunction with existing models with high recall and high false positive rates to achieve precise identification with low disruption. In practical applications, the detection of network crimes is only the beginning; it is crucial to analyze the types and causes of criminal behavior in depth, which largely depends on professional security personnel.

In terms of model inference, BT can rapidly assess each behavioral data. However, in the model training phase, the rapid evolution of cybersecurity makes it extremely challenging to maintain a model that can always perceive the latest criminal behaviors, as frequent data updates necessitate regular model updates. A simple way is to employ an incremental update mechanism, allowing the model to retain existing knowledge while absorbing new knowledge. However, this method performs poorly in mitigating feature drift, particularly in terms of new criminal behavior types, therefore, considering adopting a streaming online evolutionary strategy, maintaining a category anchor representation for criminal behavior through additional storage space, conducting momentum-based updates for optimizing anchor representations with newly appearing data, thus achieving adaptability to new behavioral data at low time and computational costs.

6.6.3 The Impact of False Negative

False negatives in cybercrime detection can potentially lead to severe consequences and have been argued to be more widely spread than street crimes. False negatives represent instances where security systems fail to correctly identify genuine threats or malicious activities, thereby exposing the system to risks posed by unrecognized threats. This section delves into the discussion of the dangers of false negatives in cybersecurity using fraudulent transactions from the OnlinePayment dataset, as well as the measures to address them.

In online finance, particularly in anti-fraud tasks, instances of missed detections are indeed more prone to occur in the context of high-value transactions. This is attributed to the elevated risks and potential for fraud associated with large transactions, making them prime targets for fraudulent actors. In our dataset, approximately 86% of undetected transactions involve amounts exceeding \$300, with about 42%

involving sums surpassing $3,000. In other security domains, such as undetected security vulnerabilities, the presence of potential vulnerabilities within the system that go unnoticed can serve as gateways for malicious hackers. Failure to identify malicious activities may escalate the risk of data breaches, jeopardizing personal privacy and confidential information. Undetected threats could lead to system breakdowns or service interruptions, impacting business continuity and stability.

In our study, we initially utilize known cybercrime cases to train behavioral models; however, some users lack transaction records in the training set and only appear in the test set, resulting in the model's inability to effectively recognize these users' behavior patterns. Statistical analysis reveals that approximately 65% of users with missed detections exhibit behavior patterns that were not learned by the model, hence imposing limitations when detecting new users. In practical deployment, we can consider implementing multi-tiered control mechanisms, such as initially employing unsupervised security measures to identify previously unseen users. Despite the typically higher false positive rates associated with unsupervised approaches, we can complement them with the precise detection provided in this study to enhance the detection capabilities of security systems and reduce the occurrence of false negatives.

6.7 Conclusion

This chapter studies how to detect hidden cybercrimes that are carefully planned. We propose the *b*ehavior *t*omographer, BT, which can project behavior structure onto slices based on different interior associations. It models expressive behavioral patterns based on sliced representations learned on reformatted slices. By learning independently on each slice, BT eliminates the superimposition of information outside the focused associations. Then it learns agent representations from attribute to event to agent in a slice-by-slice structure to achieve sliced representation fusion. Comprehensive experiments on the representative cybercrime scenarios demonstrate that our method significantly outperforms the competing methods and can more effectively detect hidden cybercrimes under the condition of a low false positive rate.

As future work, we aim to extend our work to encompass a broader range of cybersecurity scenarios, such as malicious behaviors in decentralized finance. Meanwhile, we will consider the development of specific technologies capable of effectively addressing challenges related to adversarial attacks.

References

1. D. Wang, B. Wu, X. Yuan, L. Wu, Y. Zhou, H. Cui, IEEE Trans. Serv. Comput. **Early Access**, 1 (2024)
2. S. Xi, S. Yang, X. Xiao, Y. Yao, Y. Xiong, F. Xu, H. Wang, P. Gao, Z. Liu, F. Xu, J. Lu, in *Proceedings o f the CCS 2019, London, UK, November 11-15, 2019* (2019), pp. 2421–2436
3. M. of Public Security of the People's Republic of China, The ministry of public security announced the top ten most common types of telecommunications and internet fraud. http://www.mps.gov.cn:8080/n2253534/n2253535/c9077804/content.html
4. D. Wang, H. Feng, S. Wu, Y. Zhou, L. Wu, X. Yuan, in *Proceedings of the RAID 2022, Limassol, Cyprus, October 26–28, 2022* (2022), pp. 99–114
5. S.R. Hussain, I. Karim, A.A. Ishtiaq, O. Chowdhury, E. Bertino, in *Proceedings of the CCS 2021, Virtual Event, Republic of Korea, November 15–19, 2021* (2021), pp. 1082–1099
6. D. Zhang, S. Wang, Q. Zhang, Y. Zhang, IEEE Trans. Serv. Comput. **16**(6), 4168 (2023)
7. D. Li, D. Chen, B. Jin, L. Shi, J. Goh, S. Ng, in *Proceedings of the ICANN 2019, Munich, Germany, September 17–19, 2019*, vol. 11730 (2019), pp. 703–716
8. J. Liang, W. Guo, T. Luo, V.G. Honavar, G. Wang, X. Xing, in *Proceedings of the NDSS 2021, Virtual Event, February 21–25, 2021* (2021)
9. A. Deng, B. Hooi, in *Proceedings of the AAAI 2021, Virtual Event, February 2–9, 2021* (2021), pp. 4027–4035
10. P. Marteau, IEEE Trans. Inf. Forensics Secur. **16**, 2157 (2021)
11. D. Formby, R. Beyah, IEEE Trans. Inf. Forensics Secur. **15**, 1455 (2020)
12. C. Feng, P. Tian, in *Proceedings of the ACM SIGKDD 2021, Virtual Event, Singapore, August 14–18, 2021* (2021)
13. Y. Mirsky, T. Doitshman, Y. Elovici, A. Shabtai, in *Proceedings of the NDSS 2018, San Diego, California, USA, February 18–21, 2018* (2018)
14. C. Zhou, R.C. Paffenroth, in *Proceedings of the ACM SIGKDD 2017, Halifax, NS, Canada, August 13–17, 2017* (2017), pp. 665–674
15. H. Zenati, M. Romain, C. Foo, B. Lecouat, V. Chandrasekhar, in *Proceedings of the IEEE ICDM 2018, Singapore, November 17–20, 2018* (2018), pp. 727–736
16. G. Naitzat, A. Zhitnikov, L. Lim, J. Mach. Learn. Res. **21**, 184:1 (2020)
17. M. Wang, Z. Cang, G. Wei, Nat. Mach. Intell. **2**(2), 116 (2020)
18. C. Fu, Q. Li, K. Xu, in *Proceedings of the NDSS 2023, San Diego, California, USA, February 27–March 3, 2023* (2023)
19. L. Tong, B. Li, C. Hajaj, C. Xiao, N. Zhang, Y. Vorobeychik, in *Proceedings of the USENIX 2019, Santa Clara, CA, USA, August 14–16, 2019* (2019), pp. 285–302
20. C. Wang, H. Zhu, IEEE Trans. Dependable Secure Comput. **19**(1), 301 (2022)
21. D. Tang, S. Wang, B. Liu, W. Jin, J. Zhang, IEEE Trans. Serv. Comput. **16**(5), 3373 (2023)
22. H. Zhu, C. Wang, S. Chai, IEEE Trans. Serv. Comput., **5**, 2180–2193 (2024)
23. D. Wang, S. Wu, Z. Lin, L. Wu, X. Yuan, Y. Zhou, H. Wang, K. Ren, in *Proceedings of the SBC@AsiaCCS 2021, Virtual Event / Hong Kong, SAR, China, June 2021* (2021), pp. 23–28
24. S.M. Milajerdi, B. Eshete, R. Gjomemo, V.N. Venkatakrishnan, in *Proceedings of the CCS 2019, London, UK, November 11–15, 2019* (2019), pp. 1813–1830
25. X. Feng, Q. Li, H. Wang, L. Sun, in *Proceedings of the USENIX Security 2018, Baltimore, MD, USA, August 15–17, 2018* (2018), pp. 327–341
26. Y. Ji, Z. Zhang, X. Tang, J. Shen, X. Zhang, G. Yang, in *Proceedings of the ACM SIGKDD 2022, Washington, DC, USA, August 14–18, 2022* (2022), pp. 687–697
27. D.B. Rawat, R. Doku, M. Garuba, IEEE Trans. Serv. Comput. **14**(6), 2055 (2021)
28. T. Chen, Y. Zhang, Z. Li, X. Luo, T. Wang, R. Cao, X. Xiao, X. Zhang, in *Proceedings of the CCS 2019, London, UK, November 11–15, 2019* (2019), pp. 1503–1520
29. B. Yan, C. Yang, C. Shi, Y. Fang, Q. Li, Y. Ye, J. Du, ACM Trans. Knowl. Discov. Data **18**(2), 1 (2023)

30. R. Fujimaki, in *Proceedings of the IEEE ICDM 2008, Pisa, Italy, December 15–19, 2008* (2008), pp. 797–802
31. F.T. Liu, K.M. Ting, Z. Zhou, ACM Trans. Knowl. Discov. Data **6**(1), 1 (2012)
32. G. Pang, L. Cao, L. Chen, D. Lian, H. Liu, in *Proceedings of the AAAI 2018, New Orleans, Louisiana, USA, February 2–7, 2018* (2018), pp. 3892–3899
33. W. Lin, S. Ke, C. Tsai, Knowl.-Based Syst. **78**, 13 (2015)
34. S. Han, X. Hu, H. Huang, M. Jiang, Y. Zhao, in *Proceedings of the NeurIPS 2022, Virtual Event, November 28–December 9, 2022*, vol. 35 (2022), pp. 32,142–32,159
35. Y. Yang, R. Yang, Y. Li, K. Cui, Z. Yang, Y. Wang, J. Xu, H. Xie, ACM Trans. Web **17**(3), 15:1 (2023)
36. S.Ö. Arik, T. Pfister, in *Proceedings of the AAAI 2021, Virtual Event, February 2–9, 2021* (2021), pp. 6679–6687
37. J. Wu, J. Guo, B. Hooi, in *Proceedings of the ACM SIGKDD 2024, Barcelona, Spain, August 25–29, 2024* (2024), pp. 3367–3378
38. Z. Li, Y. Zhao, J. Han, Y. Su, R. Jiao, X. Wen, D. Pei, in *Proceedings of the ACM SIGKDD 2021, Virtual Event, Singapore, August 14–18, 2021* (2021), pp. 3220–3230
39. L. Qu, H. Zhu, R. Zheng, Y. Shi, H. Yin, in *Proceedings of the ACM SIGKDD 2021, Virtual Event, Singapore, August 14–18, 2021*, (2021), pp. 1390–1398
40. X. Li, S. Liu, Z. Li, X. Han, C. Shi, B. Hooi, H. Huang, X. Cheng, in *Proceedings of the AAAI 2020, New York, NY, USA, February 7–12, 2020* (2020), pp. 4731–4738
41. J. Wang, R. Wen, C. Wu, Y. Huang, J. Xion, in *Proceedings of the WWW 2019, San Francisco, CA, USA, May 13–17, 2019* (2019), pp. 310–316
42. Z. Liu, Y. Dou, P.S. Yu, Y. Deng, H. Peng, in *Proceedings of the ACM SIGIR 2020, Virtual Event, China, July 25–30, 2020* (2020), pp. 1569–1572
43. C.L. Tan, K.L. Chiew, K.S. Yong, J. Abdullah, Y. Sebastian, et al., Comput. Secur. **95**, 101793 (2020)
44. L. Zheng, Z. Li, J. Li, Z. Li, J. Gao, in *Proceedings of the IJCAI 2019, Macao, China, August 10–16, 2019* (2019), pp. 4419–4425
45. P. Velickovic, G. Cucurull, A. Casanova, A. Romero, P. Liò, Y. Bengio, in *Proceedings of the ICLR 2018, Vancouver, BC, Canada, April 30–May 3, 2018* (2018), pp. 1–12
46. Z. Wang, W. Wei, G. Cong, X. Li, X. Mao, M. Qiu, in *Proceedings of the ACM SIGIR 2020, Virtual Event, China, July 25–30, 2020* (2020), pp. 169–178
47. J. Zeng, Z.L. Chua, Y. Chen, K. Ji, Z. Liang, J. Mao, in *Proceedings of the NDSS 2021, Virtual Event, February 21–25, 2021* (2021)
48. C. Wang, H. Zhu, IEEE Trans. Inf. Forensics Secur. **17**, 2703 (2022)
49. B. Hu, Z. Zhang, C. Shi, J. Zhou, X. Li, Y. Qi, in *Proceedings of the AAAI 2019, Honolulu, Hawaii, USA, January 27–February 1, 2019* (2019), pp. 946–953
50. C. Fu, Q. Li, M. Shen, K. Xu, in *Proceedings of the CCS 2021, Virtual Event, Republic of Korea, November 15–19, 2021* (2021), pp. 3431–3446
51. A. Pektas, T. Acarman, Soft Comput. **24**(2), 1027 (2020)
52. D.P. Kingma, J. Ba, in *Proceedings of the ICLR 2015, San Diego, CA, USA, May 7–9, 2015* (2015), p. Poster
53. Z. Hu, Y. Dong, K. Wang, Y. Sun, in *Proceedings of the WWW 2020, Taipei, China, April 20–24, 2020* (2020), pp. 2704–2710
54. A.H. Lashkari, G. Kaur, A. Rahali, in *Proceedings of the ICCNS 2020, Tokyo, Japan, November 27–29, 2020* (2020), pp. 1–13
55. W. Bialek, Proc. Natl. Acad. Sci. **119**(18), e2021860119 (2022)
56. Z. Liu, C. Chen, X. Yang, J. Zhou, X. Li, L. Song, in *Proceedings of the ACM CIKM 2018, Torino, Italy, October 22–26, 2018* (2018), pp. 2077–2085
57. Y. Zhang, Y. Fan, Y. Ye, L. Zhao, C. Shi, in *Proceedings of the ACM CIKM 2019, Beijing, China, November 3–7, 2019* (2019), pp. 549–558
58. L.O. Prokhorenkova, G. Gusev, A. Vorobev, A.V. Dorogush, A. Gulin, in *Proceedings of the NeurIPS 2018, December 3–8, 2018, Montréal, Canada* (2018), pp. 6639–6649

59. G. Somepalli, M. Goldblum, A. Schwarzschild, C.B. Bruss, T. Goldstein, *Proceedings of the NeurIPS 2022 First Table Representation Workshop* (2022)
60. G. Pang, C. Shen, A. van den Hengel, in *Proceedings of the ACM SIGKDD 2019, Anchorage, AK, USA, August 4–8, 2019* (2019), pp. 353–362
61. T. Fawcett, Pattern Recogn. Lett. **27**(8), 861 (2006)
62. C.D. Manning, P. Raghavan, H. Schütze, *Introduction to Information Retrieval*, vol. 39 (Cambridge University, Cambridge, 2008)
63. T. Lin, P. Goyal, R.B. Girshick, K. He, P. Dollár, in *Proceedings of the IEEE ICCV 2017, Venice, Italy, October 22–29, 2017* (2017), pp. 2999–3007
64. Z. Yang, Q. Xu, S. Bao, Y. He, X. Cao, Q. Huang, in *Proceedings of the ICML 2021, Virtual Event, 18–24 July 2021*, vol. 139 (2021), pp. 11,820–11,829
65. P.E. Rauber, A.X. Falcão, A.C. Telea, in *Eurographics Conference on Visualization* (2016), pp. 73–77

Chapter 7
Predicting Extraordinary Behavior Sequence by Spatiotemporal Transformer

Abstract Traffic prediction acts as a fundamental function in the management and optimization of networks and services. There are emerging requirements for extraordinary traffic prediction, including variable-term traffic series and comprehensive traffic behavior. Compared to ordinary traffic prediction, these demands call for solutions to mine rich information and predict business load under broader conditions. This chapter proposes X-Trafformer, a graph spatiotemporal transformer model, for extraordinary traffic prediction. Unlike conventional techniques that model traffic sequences, this work transforms traffic sequences into traffic behaviors under generalized objects, where behaviors are initiated by generalized objects and possess specific behavioral attributes. It allows for the prediction of multiple variables in traffic data across different networks and services, leveraging the matching of behavioral attributes among behavior objects and events. X-Trafformer incorporates multiple interrelated graph structures to capture fine-grained attribute spatiotemporal associations and coarse-grained object spatiotemporal distributions, forming the foundation for accurate prediction. Evaluation on real and representative traffic scenarios (communication traffic from Italian Telecom and business traffic from Tmall) demonstrates X-Trafformer's exceptional prediction performance at low computational costs.

7.1 Introduction to Spatiotemporal Transformer

The advancements in intelligent technologies have fueled the rapid growth of data in recent years, ushering in the era of big data [1, 2]. For instance, by 2023, mobile traffic is projected to occupy a significant portion, estimated at 58.33% [3], of the total Internet traffic. To cater to the diverse demands of users in networks and services, both academia and industry are actively exploring the integration of artificial intelligence (AI) into the development of subsequent technologies (like fifth-generation mobile network communication). The integration of AI enables networks and services to optimize themselves, leading to improved efficiency, stable network connections, and enhanced user experiences [4]. Traffic prediction, as the basic function of managing and optimizing the system of networks and services, is

C. Wang, *Computational Structural Behavior*,
https://doi.org/10.1007/978-981-95-1013-9_7

an indispensable step toward achieving intelligent management with the help of AI [5]. It can guide infrastructure to organize traffic wisely and alleviate network or service congestion caused by unexpected traffic loads [6].

In traffic prediction, ordinary research primarily focuses on forecasting future traffic conditions within networks by leveraging sequential historical traffic states [4, 5, 7, 8]. However, *extraordinary* traffic prediction for networks and services faces new challenges. One of the necessary aspects in extraordinary traffic prediction is the ability to predict the variable terms within traffic sequences. This goes beyond solely relying on the spatiotemporal information inherent in traffic data. Instead, it tends to delve deep into unraveling the intricate associations within the traffic data. Furthermore, the objects of prediction extend beyond just the infrastructure load within the network, encompassing the business load within services under broader conditions. Obviously, these two challenges pose significant difficulties for existing traffic prediction models that primarily rely on the spatiotemporal associations found within traffic data.

Currently, existing research [9–11] in ordinary traffic prediction mainly focuses on analyzing the traffic data itself but often neglect the inclusion of various factors such as base station information and correlations among variables. Some studies [12–15] have attempted to address this by integrating convolutional and recurrent operations to handle the spatiotemporal characteristics of the traffic data. However, these studies rely on sparse grid-based euclidean structures, which fail to capture the comprehensive spatial information in network topology. To overcome these limitations, there is increasing interest in leveraging non-euclidean graph structures, specifically graph representation by graph neural networks (GNNs), for spatiotemporal cellular traffic prediction [5]. On the other hand, previous studies [16–18] focusing on multivariate time series (MTS) have considered the correlations between multiple variables. But they often rely on fully connected networks to map these variables into the same space [19]. These methods are constrained by the computationally intensive learning process, particularly in practical application scenarios that involve handling large volumes of traffic data. Existing above methods aim to achieve accurate predictions by improving the utilization of information and have shown some effectiveness in ordinary traffic prediction tasks.

Faced with extraordinary challenges, efficiently achieving accurate traffic prediction remains a daunting task for the research community [20–23]. Traditional MTS-based methods treat variables equally in traffic data without considering the concept of objects, making it difficult to handle complex service scenarios [19]. For instance, in the context of grid location in the telecommunications industry, grid information is often treated as additional variables without further considering the potential preference associations among grids under different objects. On the other hand, graph-based approaches introduce the concept of objects and transform traffic data into graphs, leveraging graph-related methods for efficient information utilization [14, 22, 24, 25]. However, these approaches are only efficient in handling fixed types of objects and heavily rely on explicit associations between objects, overlooking the correlation between variables. Therefore, existing methods only partially address the challenges of nontrivial traffic prediction from specific perspectives and lack efficiency and generality in handling various types of prediction tasks in networks and services.

This chapter formulates the problem of extraordinary traffic prediction as a behavior prediction task for the first time, proposing an efficient and universal solution applicable across networks and services. Traffic data is transformed into structured tuples consisting of behavioral objects, events, and attributes, representing how behavioral objects initiate events with multiple associated attributes. Through data partitioning, coarse-grained behavioral object associations and fine-grained attribute associations are extracted, establishing the foundation for accurate predictions. Complex multivariable correlations are effectively captured via information sharing and propagation among behavioral attributes. Multivariable traffic prediction is achieved through behavioral attribute matching based on objects and events. The framework generalizes behavioral objects beyond specific business entities to any variable, enabling diverse traffic data across networks and services to share behavioral semantics in a unified representation space. This enhances the relevance between objects and attributes. For instance, when predicting traffic for a specific seller, the seller serves as the object; when predicting product traffic, the seller ID becomes an attribute. Exploring relationships between the seller ID and other attributes improves product modeling for more accurate predictions.

This chapter proposes the X-Trafformer framework (as depicted in Fig. 7.1) for extraordinary traffic prediction. It defines the behavior objects from the traffic data and decomposes the traffic data into several behavioral attributes, where the behavior object can be generalized to any term rather than traditional wireless infrastructure. Information learning among attributes and objects separately will enhance the behavior layer for traffic prediction. It focuses on modeling for behavior objects, behavior events, and behavioral attributes and aims to capture spatiotemporal associations at fine-grained multivariable levels and spatiotemporal distributions at coarse-grained object levels. The framework consists of three modules: object layer, event layer, and attribute layer. The object layer incorporates various types of behavioral object information, including spatial attributes such

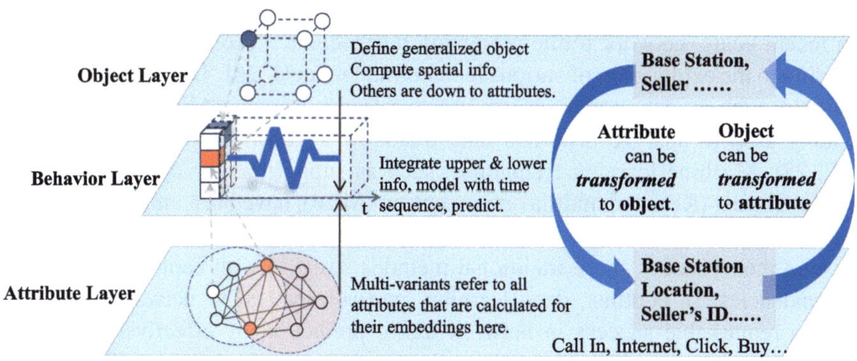

Fig. 7.1 Illustration of behaviorizing traffic data

as node position, topological distance, and communication intensity. In scenarios where explicit object information is unavailable, the framework introduces the use of dynamic time warping (DTW) matrices. The attribute layer is designed to capture correlations among multiple variables using association mapping and graph structures. The event layer incorporates advanced time series models, aligning behavioral object information and behavioral attribute information through joint representation module. This ensures consistency during the data integration process and prevents domination by static variables. By selectively incorporating the spatiotemporal features of behavioral objects and correlations among variables into behaviors using an attention mechanism, our framework significantly improves the accuracy of traffic prediction.

The main contributions can be summarized as follows:

- This work proposes the X-Trafformer model, which innovatively transforms traffic prediction into the prediction of behaviors. This model converts traffic sequences into multilayered behavior graphs, effectively addressing the challenge of extraordinary traffic prediction with limited data by integrating multiple granularity levels of behavioral information.
- This work designs a joint representation module that utilizes an attention-based information integration mechanism to efficiently align and integrate the captured spatiotemporal associations at fine-grained variable-term levels and the spatiotemporal distribution at coarse-grained object levels. It improves the scalability and computational efficiency significantly.
- This work is evaluated on real traffic datasets from two representative network/service scenarios. The experimental results show that X-Trafformer performs outstandingly on multiple typical metrics and surpasses SOTA models. Further analysis reveals the work's computational efficiency, delivering high prediction performance while maintaining low computational and time costs.

7.2 From Traffic Prediction to Behavior Prediction

In recent years, network traffic prediction has become a prominent research area, leading to the exploration of various methods and models [26]. Traditional statistical models, such as autoregressive integrated moving average (ARIMA), have been widely used to analyze historical data trends to forecast future traffic variations [27, 28]. Machine learning techniques, such as support vector machines (SVM), random forest (RF), and multilayer perceptron (MLP), have been employed to learn patterns and correlations in data, thereby boosting prediction accuracy. However, despite their success, these traditional methods struggle with capturing long-term dependencies. In contrast, deep learning, particularly LSTM networks [29], has shown remarkable success in network traffic prediction by effectively capturing sequential patterns and long-term dependencies [10, 30]. Furthermore, the adoption of multitask learning and cross-domain datasets enhanced prediction accuracy and allowed predictions across various network units or domains.

The aforementioned methods mainly focus on capturing the temporal characteristics of traffic data, neglecting the mutual influence among neighboring network nodes. For instance, the traffic variations between adjacent base stations are highly associated, and considering the associations of objects can enhance the predictive performance. To address this limitation, some works [7, 31] utilized convolutional neural networks (CNN) or graph convolutional neural networks (GCN) [32] to extract spatial features from the data and improve prediction accuracy. Spatiotemporal graphs are widely employed as a data structure for representing network traffic in traffic prediction tasks [13, 22, 33]. It enables researchers to analyze the associations between traffic data at different locations and leverage the associations among nodes in the graph to enhance predictive accuracy. DCRNN [14] utilized diffusion convolution for modeling and extracting spatial features, which has shown outstanding performance in capturing the spatiotemporal characteristics of traffic. However, it is worth noting that spatiotemporal graph-based methods tend to prioritize spatial features, potentially resulting in limited computational resources to simultaneously emphasize the temporal characteristics of the sequences themselves, particularly for sequences with pronounced periodicity. It may result in the loss of crucial information relevant to traffic prediction.

Another research direction in traffic prediction focuses on treating it as a multivariate time series (MTS) prediction task, incorporating correlations among multiple relevant variables to enhance the comprehensiveness and accuracy of traffic prediction. Transformer-based methods [8, 34] have emerged as significant approaches in this domain, leveraging their powerful modeling capabilities and parallel processing capacity to capture complex dependencies across various dimensions and time scales. Several variations of the transformer architecture [35] have been proposed for traffic prediction. For example, Informer [19] introduced ProbSparse self attention, Autoformer [23] incorporated an automatic relevance mechanism, and Crossformer [36] employed a two-stage attention mechanism. These variants refine the transformer model from different perspectives and have demonstrated commendable performance in traffic prediction tasks.

Transformer-based methods have certain limitations. Firstly, these methods lack graph learning modules, which may result in less effective consideration of inter-sequence correlations compared to spatiotemporal graph-based methods. Secondly, incorporating inter-sequence correlations would require additional modules or customized modifications to the attention mechanism (as seen in approaches like Crossformer). Given the high computational demands of the transformer architecture, introducing spatial association modules would lead to a significant increase in the computational burden. Therefore, some existing approaches often mitigate this issue by reducing the number of sequences or employing partitioning prediction strategies. Additionally, researchers have explored collaborative optimization across different end devices to reduce resource consumption [37, 38]. For instance, Zhou et al. proposed an energy-efficient virtual machine allocation and deployment algorithm based on an adaptive energy-aware framework, ensuring high quality of service while reducing energy consumption [39]. In contrast, our proposed approach circumvents underlying modifications to the existing architecture. Instead, we introduce the concept and attribute learning independently and then efficiently perceive this information by aligning and fusing it, achieving efficient computation.

7.3 Overview

7.3.1 Architecture

We innovatively transform the task of traffic prediction into behavior prediction and propose the X-Trafformer framework specifically designed for this purpose. Leveraging traffic data itself, we introduce additional behavioral object correlations and fine-grained behavioral attribute information to enhance high-performance predictions in network services. Our framework consists of three main modules: the attribute layer, the object layer, and the behavior layer (as illustrated in Fig. 7.2).

The object layer considers various types of object information, including node spatial positions, topological distances, and communication intensity. In cases where explicit spatial information is unavailable, we employ dynamic time warping (DTW) matrices. The attribute layer utilizes association mapping and graph structures to capture correlations among multiple behavioral attributes. The behavior layer includes a joint representation module to align objects and attribute information, ensuring consistency during data integration and mitigating the dominant role of static variables. Advanced time series models are employed for learning and predicting time sequences.

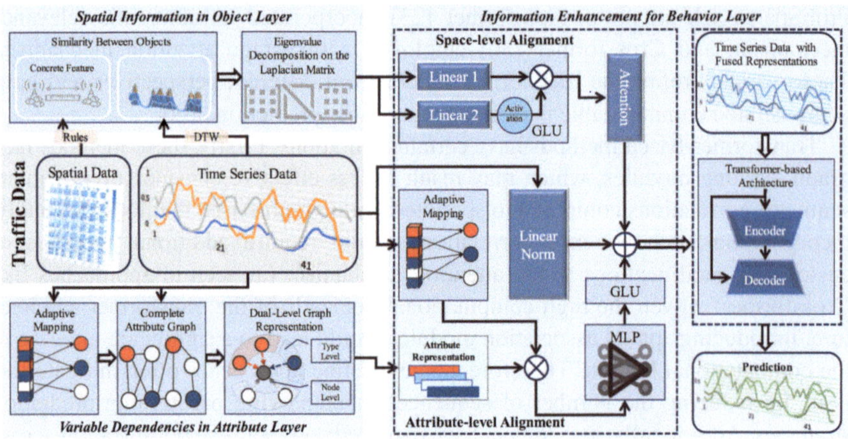

Fig. 7.2 The framework of X-Trafformer

The rationale behind our use of a hierarchical architecture is to significantly reduce complexity, drawing inspiration from the concept of deep convolutional neural networks. The effectiveness of a large number of layers in a small network surpasses that of a small number of layers in a large network, with parameter growth being linear rather than quadratic or exponential. Another advantage of the hierarchical structure is the ability to customize the model for each layer. As long as the final output is in the form of embeddings, a standard input sequence can be obtained through joint processing, eliminating the need to modify the final sequence model. Consequently, for the object layer and attribute layer, we can design them in a refined and customized manner to better meet the requirements of the scenario, fully exploiting the role of information mining, without concerns about subsequent computational complexity explosions or the need for input-output alignment adjustments.

7.3.2 Object Layer

The generalized object in different hierarchical levels of network services can be set to different objects. For instance, when focusing on the network layer, we can consider base stations as the objects, with the inflow and outflow of traffic as their behavioral attributes. On the other hand, when dealing with the service layer, let us take the Tmall shopping dataset as an example. If we take product categories as the object, we can obtain traffic prediction results specific to each category. Similarly, considering the sellers as the objects would yield predictions of their respective visitation traffic. This illustrates the essence of generalized objects, where objects are not fixed, and everything can be regarded as an object.

Concerning the selected objects, we investigate their coarse-grained spatial distribution. In datasets related to telecommunications, where the objects hold physical significance, the correlation among different nodes often demonstrates regional clustering, quantifiable through spatial distance. Illustrated in Fig. 7.3, we depict a heatmap portraying the communication strength between grids in the cities of Milan and Trento. Each grid is identified by a grid ID, with coordinates (x, y) representing the communication intensity. Notably, the heatmap reveals discernible clustered patterns, where adjacent grid IDs correspond to neighboring grids. This finding provides empirical support for spatial distribution patterns in inter-grid communication dynamics. However, for abstract objects without physical spatial meanings or nodes lacking spatial distribution data, we can adaptively map them to a metric space based on the correlation of multiple sequences. Through these analyses, we can categorize the metric space into concrete and abstract types.

Integrating spatial information of objects provides a comprehensive understanding of the network's interconnectivity and patterns. Directly inputting node spatial coordinates would increase computational complexity and risk overfitting. To address this, we draw inspiration from graph representation learning techniques [40, 41] and present a non-euclidean representation learning approach based on the

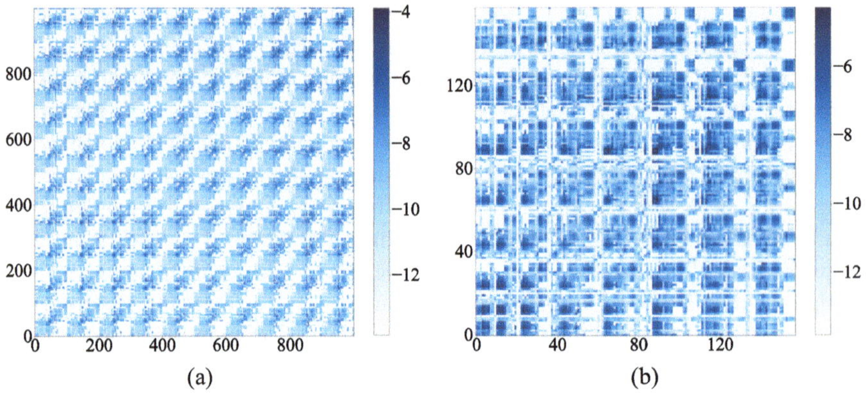

Fig. 7.3 The heatmap of the interaction strength between nodes in each region in Milano and Trento. (**a**) Milano. (**b**) Trento

eigendecomposition of the Laplacian matrix. This method transforms the network structure into a vector space representation to capture the mutual dependencies between nodes.

7.3.3 Attribute Layer

The attribute layer aims to address the limitations of existing transformer-based methods for MTS in extraordinary traffic forecasting tasks. Conventional methods directly input multivariable sequences as features into transformer models. However, in traffic prediction, the increasing number of variables makes it challenging for these models to effectively capture correlations. The limited receptive field of convolutional layers also hinders the learning of global associations among variables. As an alternative solution, utilizing linear layers requires constructing a parameter matrix that is positively correlated with the number of variables and the value space of attributes, which significantly increases training complexity with suboptimal results.

To address these challenges, we propose a hierarchical approach that allows us to capture attribute information without introducing excessive computational costs. We treat the attribute layer as a distinct module, acquire representations for attribute values, and subsequently feed them into the subsequent behavior layer for learning behavior sequences. By separating the attribute layer from the behavior layer, we can better capture the relationships between attributes and behaviors while managing computational complexity.

In the attribute layer, we delve deeper into the correlations among the behavioral attributes of each object. Our method involves constructing a complete attribute graph prior to variable processing, enabling us to model the interdependencies and correlations among different variables from a global perspective. By employing a complete attribute graph, we overcome the limitations of local receptive fields and capture complex associations and dependencies among variables.

In the complete attribute graph, nodes represent variables, and edges indicate the strength of correlation or similarity between variables. We leverage graph learning algorithms to obtain global node representation for each variable. By utilizing graph neural networks with dual-level attention mechanisms to handle the complete attribute graph, our model adaptively aggregates information from neighboring nodes and propagates it to the target node. Through multiple rounds of graph learning operations, we gradually extract higher-level features and interdependencies among variables. It allows us to effectively capture fine-grained correlations between behavioral attributes, enhancing the model's capability to handle complex traffic forecasting scenarios. By considering the global perspective and leveraging graph learning techniques, our attribute layer enables a more comprehensive understanding of the relationships among variables, leading to improved forecasting performance in complex traffic scenarios.

7.3.4 Behavior Layer

We consider the traffic sequence as a series of traffic behaviors, which incorporates object and attribute information from the previous two layers. In this layer, we aim to learn the pattern within these behavior sequences and make accurate predictions. To achieve this, we need to integrate the object representation and attribute representation, which have different dimensions and prevent direct integration. Therefore, we employ a dedicated module to map both the object and attribute information into the same space as the time series data. This module consists of key components:

Spatial Alignment Spatial alignment is crucial because the attribute and object information may have different spatial representations. It ensures that similar attribute and object features have consistent representations in the corresponding time series data. Techniques such as temporal or spatial alignment methods can be used to achieve this, establishing a coherent spatial association across the data.

Spatiotemporal Information Learning Once the dimensions are aligned, the next step is to fuse these features with the time series data. This fusion process can be accomplished through simple concatenation operations or more sophisticated feature fusion models. This integration enables the seamless combination of attribute, spatial, and temporal information, which is crucial for capturing the complex patterns in the behavior sequences. Once we have obtained the behavioral sequence with spatiotemporal information, we can approach the problem as forecasting the future values of individual time series based on their past observations. This can be treated as a multivariate time series data analysis.

7.4 Graph-Based Spatiotemporal Transformer

Before describing the detailed methods, we summarize the notations of key parameters in Table 7.1 as preparations.

Problem Definition. Let $\mathscr{S} = \{S_1, S_2, ..., S_L\}$ represent a time series dataset of length L. This dataset can be either network traffic data or business data from services. Each data point $Si = \{a_1, a_2, ..., a_C\}$ consists of a set of C variables. Our objective is to leverage the spatial information of the objects generating the time series data and the temporal information in $\mathscr{S}$ to learn a function that can accurately predict the future values of each variable in $\mathscr{S}$. By integrating the spatial information of the object and the temporal patterns in the data, we aim to improve the accuracy of the predictions for each variable in $\mathscr{S}$.

7.4.1 Decomposition-Based Representation for Object Layer

We propose a novel non-euclidean representation learning method that enables unsupervised learning of spatial representations for nodes using their adjacency information. Unlike traditional methods that solely rely on euclidean distance information, our method captures topological information between nodes, enriching the learned representations.

We begin by constructing a graph G_O that represents the adjacency relationships between objects, where N_O is the node set of objects. In this graph, each node n corresponds to a specific object, where $n \in N_O$, and edges represent the connections or adjacency relationships between nodes. The graph can be either undirected or directed, depending on the scenario. The Laplacian matrix is a crucial mathematical tool used to describe the structure of the graph. It helps us analyze the properties of the graph and its connectivity. Given the graph G_A, we proceed

Table 7.1 Notations of key parameters

Variable	Description
$\mathscr{S}$	The set of all time series data
S_{o_i}	The time series of object o_i
$\mathscr{O}$	The object set
$\mathscr{A}$	The attribute values set
G_O	The graph of objects
N_O	The node set of objects
G_A	The graph of attributes
N_A	The node set of attributes
ψ	The mapping from $\mathscr{A}$ to N_A
h	The hidden representation in neural networks
σ	The activation function
dim	The dimensionality of representations

to construct the Laplacian matrix. For an undirected graph, the Laplacian matrix is obtained as the difference between the degree matrix and the adjacency matrix: $L_{G_O} = D_{G_O} - Adj_{G_O}$, where D_{G_O} is the diagonal degree matrix and Adj_{G_O} is the adjacency matrix. In the case of a directed graph, a weighted matrix is used instead, resulting in $L_{G_O} = D_{G_O} - Wt_{G_O}$, where Wt_{G_O} is the weighted matrix.

7.4.1.1 Similarity Between Objects

To reflect the strength of connections between the object nodes, we propose two features to calculate the similarity between any two object nodes. These features are used to construct the weighted matrix Wt_{G_O} in order to capture the connectivity strength between nodes.

Concrete Spatial Features Specific spatial features include, but are not limited to, the following, depending on availability in a specific scenario.

Physical Distance. Utilizing node coordinates, we can calculate the euclidean distance or other distance metrics between them. Nodes in close physical proximity are likely to have higher associations, leading to more frequent communication and resource sharing.

Route Distance (Hop Count). The route distance between nodes can be represented by the hop count, indicating the number of intermediate nodes required to establish a connection between them. A lower hop count signifies a shorter path and stronger interrelation between the nodes.

Communication Strength. The interaction intensity between nodes can be quantified based on factors such as communication frequency, data exchange volume, or joint participation in shared activities. Nodes with higher communication strength exhibit stronger interrelations.

These concrete spatial features provide valuable insights into the spatial distribution and connectivity patterns within the network, offering a basis for analyzing network dynamics and predicting node behaviors. It is worth noting that concrete spatial features are not necessary. In many scenarios and datasets, meaningful physical information is difficult to obtain.

Abstract Spatial Features In multivariate time series (MTS) prediction, we employ dynamic time warping (DTW) to compute the distance or similarity between each pair of sequences (S_{o_i}, S_{o_j}). Here, S_{o_i} and S_{o_j} denote the sequences generated by objects o_i and o_j, respectively. By computing the DTW distance between multiple pairs of sequences, we can obtain a distance matrix or similarity matrix that represents the spatial correlation between the sequences [5]. The specific calculation formula is as follows:

$$
\begin{aligned}
DTW(o_i, o_j) =& Distance(o_i, o_j) + \\
& \min\{DTW(o_{i-1}, o_j), DTW(o_i, o_{j-1}), \\
& DTW(o_{i-1}, o_{j-1})\},
\end{aligned}
$$

$$
Distance(o_i, o_j) = (1/S_{o_i}) * \sum |S_{o_i} - S_{o_j}|,
$$

where o_i and o_j represent two distinct objects and $|S_{o_i}|$ represents the length of the sequences S_{o_i}.

7.4.1.2 Eigenvalue Decomposition on the Laplacian Matrix

Based on the two types of features, we can construct the weight matrix Wt_G that represents the similarity between objects. Each element of the weight matrix represents the weight between two objects, which can be defined based on the concrete spatial features and abstract spatial features. For example, we can map the two kinds of features to a range of [0, 1] and then take the average.

Subsequently, we perform eigenvalue decomposition on the Laplacian matrix L_{G_O}. This process factorizes L_{G_O} into eigenvalues and their corresponding eigenvectors, i.e., $L_{G_O} = \mathscr{E}_{G_O} \Lambda \mathscr{E}_{G_O}^T$, where $\mathscr{E}_{G_O}$ is the eigenvector matrix, and Λ is the diagonal eigenvalue matrix. Here, each row of the eigenvector matrix E_{G_O} can be regarded as the object representation of the corresponding object o_i, denoted as $\vec{o}_i$. During the eigenvalue decomposition process, the eigenvector matrix $\mathscr{E}_{G_O}$ enables us to capture spatial information. The eigenvectors correspond to distinct spatial patterns, representing the distribution of traffic across different objects. Larger eigenvalues signify more significant spatial patterns, which allow us to identify key factors for network traffic prediction.

7.4.2 Graph-Level Representation for Attribute Layer

We denote a complete attribute graph (CAG) as G_A, where nodes represent the values of variables and edges represent the number of connections between variable values, indicating the strength of their associations. Given all multivariate time series data $\mathscr{S} = \{S_1, S_2, ..., S_L\}$, each individual time series S_i generated by object o consists of multiple attributes $\{a_1, a_2, ..., a_C\}$, where L is the size of $\mathscr{S}$ and C is the number of attributes.

7.4.2.1 Adaptive Mapping Function

Our goal is to create a balanced complete attribute graph without attribute nodes clustering around a few central nodes. To achieve this, we denote the set of attribute values as $\mathscr{A}$ and the set of nodes in G_A as N_A. The set $\mathscr{A}$ can be obtained by traversing the time series data S with $C \cdot L$ operations, where $C = |S|$ and $L = |\mathscr{S}|$. We design the adaptive mapping function (AMF) to achieve $\psi(\cdot) : \mathscr{A} \rightarrow N_A$ that assigns each attribute value in time series data to distinct nodes in G_A.

To ensure the equal number of attribute value mappings for each node, we follow the following ways. First, we perform a quick sorting of the attribute values in $\mathscr{A}$, with a time complexity of $O(|\mathscr{A}| \log |\mathscr{A}|)$. Because $|\mathscr{A}| \leq C \cdot L$, its time complexity is therefore equivalent to $O(C \cdot L \cdot \log(C \cdot L))$. Then, we divide the sorted attribute values into K equal-sized partitions denoted as $\mathscr{A}^1, \mathscr{A}^2, \ldots, \mathscr{A}^K$, where K is the predefined hyperparameter and each partition $\mathscr{A}^i$ contains an equal number of attribute values. K controls the granularity of the behavioral attribute values, where a larger K implies fewer behavioral attributes being assigned to a single node within a range. For example, if a count-based attribute has values ranging from 1 to 1000, we can set K to 100 to map the 1000 data values onto 100 nodes. In this case, ten similar attribute values are considered as one node in G_A. These ten attributes share the same contextual and structural information within G_A. Subsequently, we assign the attribute values from the same partition $\mathscr{A}^i$ to a corresponding node in N_A.

By adopting this strategy, we aim to achieve a balanced distribution of attribute value mappings across nodes. This ensures that the constructed CAG captures global associations among attributes in an unbiased manner, without being influenced by a few central nodes.

7.4.2.2 Constructing Complete Attribute Graph

In G_A, our focus is on capturing the correlations between variables. By applying an adaptive mapping function to all the time series data, we obtain the mapping $\psi(\cdot)$ from attributes to nodes, as well as the set of nodes N_A. We consider N_A as the node set in G_A. The next step is to determine how to convert the correlations between variables into edges between nodes.

The co-occurrence of two attributes in a time series data naturally indicates their correlation. To achieve this, we iterate through each data S in the set $\mathscr{S}$. For any pair of attribute values (a_l, a_j), we add their co-occurrence relationship to E_A, which represents the edge set of G_A. In a data S, we can generate C^2 attribute pairs, so we need to perform $C^2 \cdot L$ operations for the entire dataset $\mathscr{S}$. It is important to note that a_i and a_j are attribute values, not nodes in N_A. Therefore, we use $\psi(a_i)$ and $\psi(a_j)$ to convert them into nodes, and add $(\psi(a_i), \psi(a_j), \tau)$ to E_A, where τ represents the type of relationship between the attributes. This relationship type can be composed of two attributes, such as "place-amount" for a location attribute and an amount attribute. In our work, we consider the co-occurrence relationship as an undirected association, so we simultaneously add $(\psi(a_j), \psi(a_i), \tau)$ to E_A. Our algorithm builds the CAG by iterating over the MTS data, transforming the variables using an adaptive mapping function $\psi(\cdot)$, and adding co-occurrence between different variables. The process of building a complete attribute graph is described in Algorithm 5.

After following these steps, we construct a complete attribute graph G_A. As an illustration, let us consider the Telecom-M dataset and focus on five variables: $smsin$, $smsout$, $callin$, $callout$, and $internet$. To create G_A, we map the values of these five variables to different types of nodes. Then, we establish connections between all the nodes within each data row, indicating their co-occurrence. This process is repeated for every training data instance, ensuring that we capture the relationships among the variables in the dataset.

In terms of computational cost, since C is much smaller than L, the total time complexity of constructing the complete attribute graph from the raw data can be classified as $O(L \log L)$. In our framework, we subsequently employ a transformer-based architecture for final time series prediction, such as Informer with $O(L \log L)$ complexity and Autoformer with $O(L \log L)$ complexity. Meanwhile, this operation is only deployed once before the formal training of the model, so its overall resource consumption is not significant.

7.4.2.3 Dual-Level Graph Representation

To effectively learn the representations of nodes in G_A, we introduce a dual-level attention mechanism [42, 43] that calculates node attention scores from both a type-level and a node-level perspective. This dual-level approach efficiently captures the variations in importance at different levels. This dual-level attention mechanism enhances the discriminative power of our representation learning process, as it effectively captures the varying importance of nodes in G_A at different hierarchical levels.

Type-Level Learning Neighboring nodes with the same type are aggregated. For an edge type τ, we compute the edge representation of a node $n \in N_A$, $h_n^\tau = \sum_{n'} h_{n'}$, where n' belongs to the neighbors set of n with type τ and h_n is the node representation of a node n. Then, we calculate the type-level attention score α_τ based on h_n and h_n^τ:

$$\alpha_\tau = \text{Softmax}\left(\sigma\left(\mu_\tau \cdot \left[h_n \| h_n^\tau\right]\right)\right), \tag{7.1}$$

where σ represents the activation function, and μ_τ^T is the transpose of attention vector of the type τ.

Algorithm 5: The procedure of building CAG

Input: The multivariate time series data $\mathscr{S}$, the predefined number of attribute nodes K
Output: The adjacency matrix M of CAG

Initialize an empty attribute set $\mathscr{A}$;
Time complexity $O(C \cdot L)$, $C = |S|$ and $L = |\mathscr{S}|$;
foreach $S \in \mathscr{S}$ **do**
  foreach $a_i \in S$ **do**
    Add a_i into $\mathscr{A}$;
  end
end
Perform Quicksort and $|\mathscr{A}|$ operations with $|\mathscr{A}| \leq C \cdot L$;
Time complexity $O(C \cdot L \cdot (\log(C \cdot L) + 1))$;
Function $AMF(\mathscr{A}, K)$ **is**
  Initialize $\psi(\cdot)$ as an empty set of mapping rules;
  Initialize an empty node set N_A;
  Quicksort $\mathscr{A}$ in ascending order;
  Divide $\mathscr{A}$ into K equal-sized partitions:

$$\mathscr{A}^i = \begin{cases} \mathscr{A}\left[\frac{(i-1)\cdot|\mathscr{A}|}{K} + 1 : \frac{i\cdot|\mathscr{A}|}{K}\right], & \text{if } i < K \\ \mathscr{A}\left[\frac{(K-1)\cdot|\mathscr{A}|}{K} + 1 : |\mathscr{A}|\right], & \text{if } i = K \end{cases}$$

  where $1 \leq i \leq K$;
  for *each partition* $\mathscr{A}^i$ **do**
    Generate an attribute node n_i;
    Add n_i into N_A;
    for *each attribute* a_j *in* $\mathscr{A}^i$ **do**
      Add a new mapping rule to ψ:
      $\psi(a_j) \to n_i$
    end
  end
  return $\psi(\cdot) : \mathscr{A} \to N_A$;
end
Initialize a graph $G_A = \{N_A, E_A\}$;
Adaptive mapping function $\psi(\cdot) = AMF(\mathscr{A}, K)$;
Time complexity $O(C^2 \cdot L)$;
foreach $S \in \mathscr{S}$ **do**
  foreach $a_i \in S$ **do**
    foreach $a_j \in S$ *and* $a_i \neq a_j$ **do**
      Add n_{src} and n_{des} into N_A;# $n_{src} = \psi(a_i)$, $n_{des} = \psi(a_j)$;
      Add (n_{src}, n_{des}, τ) into E_A;# τ is the edge type depending on nodes
    end
  end
end
Return G_A.

Node-Level Learning The node-level attention score $\beta_{nn'}$ is obtained by incorporating the type-level attention score α'_τ with node n' into the attention mechanism:

$$\beta_{vv'} = \text{Softmax}\left(\sigma\left(\nu^T \cdot \alpha_{\tau'}\left[h_n | h_{n'}\right]\right)\right), \tag{7.2}$$

where ν^T represents the transpose of attention vector and σ represents the activation function. In our work, we employ Leaky ReLU as the activation function.

The hidden vector propagation rule for all nodes is

$$\mathbf{H}^{(l+1)} = \sigma\left(\sum_\tau \mathbf{B}_\tau \cdot \mathbf{H}_\tau^{(l)} \cdot W_\tau^{(l)}\right), \tag{7.3}$$

where l is the network layer, $\mathbf{B}_\tau$ is the attention matrix with the element $\beta_{vv'}$ in the nth row n'th column, and W_τ is the transformation matrix in model.

7.4.3 Adaptive Information Enhancement for Behavior Layer

7.4.3.1 Attribute-Level Alignment

After performing representation learning on the CAG G_A, we obtain the representation $\vec{n}_a \in \mathbb{R}^{K \times dim_a}$ for each node in N_A. These node representations $\vec{n}_a$ capture the essential characteristics of the corresponding nodes.

To further process these node representations, we use a multilayer perceptron (MLP) to transform them into a final representation output of dimension dim_m. The MLP applies a series of nonlinear transformations to the node representations, and the output of the MLP serves as the final representation for each node in N_A. As a result, we obtain an output of dimension dim_m that encompasses a high-level representation of the input values within the attribute space. This output can be further propagated to subsequent models or tasks, such as time series modeling or prediction.

7.4.3.2 Space-Level Alignment

In the context of network traffic prediction, the geographical location of each node remains unchanged, resulting in the spatial matrix that is considered a static covariate. In the Fusformer [20] approach, the authors explored the integration of covariates into time series forecasting by aligning the dimensions of the covariates with other variables. Building upon this idea, we introduce a customized approach to align the spatial embedding matrix $\mathcal{E}_{G_O} \in \mathbb{R}^{|N_O| \times dim_s}$ (obtained in Sect. 7.4.1) with the dimension of dim_m. Here, dim_s is the dimensionality of object representations.

To incorporate the spatial matrix as a covariate, we initially apply a linear transformation to map the spatial matrix to a higher-dimensional space. This transformation enhances the expressive capacity of the spatial matrix and introduces additional feature dimensions to capture subtle geographical variations. Next, we process the covariates through a gated linear unit (GLU). The GLU consists of two main components: the gating part and the linear part. The gating part typically employs a sigmoid activation function, which generates a gating vector ranging between 0 and 1. This gating vector determines the degree to which the output of the linear part is activated. When the elements of the gating vector approach 1, the output of the linear part is preserved, while values close to 0 suppress the output. The linear part is commonly implemented as a fully connected layer, mapping the input vectors to a higher-dimensional feature representation. The gating vector is then multiplied element-wise with the output of the linear part, producing the final output of the GLU. Given the representation $\vec{o} \in \mathscr{E}_{G_O}$ of an object o, we can compute its subsequent representation using the following formula:

$$\vec{o} := (\vec{o} \times \mathbf{W_1} + \mathbf{b}) \otimes \sigma(\vec{o} \times \mathbf{W_2} + \mathbf{c}), \tag{7.4}$$

where $\mathbf{W_1}$ and $\mathbf{W_2}$ are weight matrices, $\mathbf{b}$ and $\mathbf{c}$ are bias terms, $\otimes$ denotes element-wise multiplication, and $\sigma(\cdot)$ is the sigmoid activation function.

The GLU serves two primary purposes. Firstly, the gating part allows for the selection of important input features, enabling the model to focus on the most relevant information for its predictions and reducing the impact of noisy or irrelevant features. Secondly, by incorporating the gating mechanism, the model can adjust the current features based on the contextual information from the input, facilitating the modeling of long-range dependencies and capturing complex relationships.

The object representations obtained after passing through the GLU are subsequently processed by the attention module for weighted summation. This attention mechanism enables the allocation of attention weights to different types of spatial features in a meaningful way, based on their importance. For the representations $\vec{\mathscr{O}}$ of all objects obtained from Eq. (7.4), the attention weight $att_{\mathscr{O}}$ is calculated as follows:

$$att_{\mathscr{O}} = \text{Softmax}(\mathbf{W}_a \times \vec{\mathscr{O}}), \tag{7.5}$$

where $\mathbf{W}_a$ represents the attention weight parameter matrix, and Softmax$(\cdot)$ is an activation function.

7.4.3.3 Spatiotemporal Information Learning

Once the dimensions of representations are aligned, the next step involves fusing these representations with the time series data. This fusion process can be achieved through simple concatenation operations or more advanced feature fusion models. This integration facilitates the seamless combination of attribute, spatial, and temporal information, which is crucial for capturing the intricate patterns in the behavior sequences.

By utilizing this customized alignment method, we successfully achieve the alignment of the spatial matrix. The utilization of linear transformation and attention mechanisms enhances the exploitation of spatial information, thereby contributing to improved performance and accuracy in the task of network traffic prediction.

It is important to emphasize that specific alignment methods and transformation operations can be adjusted and optimized based on the requirements of the specific problem. This tailored approach offers flexibility in handling different spatial embedding matrices and allows for appropriate adjustments to achieve optimal performance and results.

In our research, we adopt a transformer-based architecture for time series learning, which consists of an encoder and a decoder. The encoder utilizes self-attention mechanisms to process the input time series data and learn meaningful representations. Conversely, the decoder generates predictions by leveraging the encoded representations and the target time series. With the attention mechanism of the transformer, the model effectively captures the complex relationships and dependencies among variables in the multivariate time series data. Furthermore, it offers advantages in modeling long-term dependencies and accommodating variable-length sequences, making it a suitable choice for capturing the intricate temporal dynamics in traffic prediction tasks.

7.5 Evaluation of X-Trafformer

We aim to answer the following research questions:

- **Performance Comparison (RQ1)**: How much performance improvement does our framework bring?
- **Ablation Study (RQ2)**: What is the influence of different modules within the framework on performance?
- **Hyperparameter Study (RQ3)**: How do the hyperparameters affect the performance of X-Trafformer?
- **Computational Cost (RQ4)**: What is the impact of different modules on the computational complexity?
- **Case Study (RQ5)**: What insights can be gained from specific cases or applications?

7.5.1 Experimental Setup

7.5.1.1 Datasets

The *Telecom* dataset used in this chapter is derived from the *Big Data Challenge* project initiated by Telecom Italia [44]. The dataset primarily consists of call detail records from the city of Milan (*Telecom-M*) and the province of Trentino (*Telecom-T*) in Italy. Telecom records the telecommunication activities of users, with each log in the dataset containing five fields: SMSin, SMSout, Callin, Callout, and Internet. These fields reflect the level of interaction between users and the mobile network within a given grid over a specific time interval (e.g., activity proportional to the amount of sent SMSs). We use the data collected at a frequency of every 10 minutes from November 1 to November 30, 2013, spanning a duration of one month. To account for irregular spatial distribution, the dataset aggregates telecommunication activities into grids with square units, simplifying the geographical management of the data and providing insights into communication intensity across different regions. In this dataset, we treat the grid regions as objects and the aforementioned five fields as attributes.

We make the pioneering use of the *Tmall* dataset [45] in the context of traffic prediction. The Tmall dataset used is an open-source dataset that includes 1.3 billion records of user visits to product listings on the Tmall platform, along with information on 90 different product categories. In our study, we define traffic as the number of visits to different product categories within a specific time interval, such as 60 minutes. The timeframe of the behaviors we used covers from April 1, 2013, to October 1, 2013. It includes four different types of actions: click, collect, cart, and buy. We consider the product categories as objects and the aforementioned four action types as attributes.

In the original dataset, the time series data is recorded in the format $(o, timestamp, field_1, field_2, ...)$. In our study, we transform the data by taking the timestamp as the index and converting it into the form $(timestamp, field_1^{o1}, field_2^{o1}, ..., field_1^{o2}, field_2^{o2}, ...)$. This transformation allows us to predict all attributes of all objects at the same timestamp. Therefore, the dimensionality of the time series is equal to the number of objects multiplied by the number of attributes. The statistic of the two datasets can be found in Table 7.2.

To ensure fair comparison, following the majority of previous works, we divide the datasets (Telecom dataset with a 10-minute interval and Tmall dataset with a 60-minute interval) into fixed time intervals. We then scale the data using the min-max normalization method, transforming it into a range of [0, 1] as the input. The data was subsequently split into training, validation, and testing sets in a proportion of 7:1:2. All experiments are repeated 5 times to calculate average results and are conducted on a remote server equipped with dual Tesla V100 32GB GPUs.

Table 7.2 Statistics of datasets. N# refers to number, while Dim# refers to the dimensionality of the data

		N# of			Dim# of
Dataset	Interval	series	object	attribute	series
Telecom-M	10min	4,176	1,000	5	5,000
Telecom-T	10min	4,176	158	5	790
Tmall	60min	4,393	90	4	360

7.5.1.2 Metrics

To evaluate the predictive performance, we employ two widely used regression metrics: mean squared error (MSE) and mean absolute error (MAE).

$$\text{MAE} = \frac{\sum_{t=1}^{T} \mid \hat{y}_t - y_t \mid}{T}, \ \text{MSE} = \frac{\sum_{t=1}^{T} \left(\hat{y}_t - y_t\right)^2}{T},$$

where $\hat{y}_t$ is the predicted value at time step t and y_t is the corresponding ground-truth value. The summation is performed over all time steps T in the dataset. MAE is robust to outliers and does not overly penalize large errors. MSE is sensitive to outliers and penalizes larger errors more heavily. It provides a measure of the overall variability or dispersion of the errors.

7.5.1.3 Baselines

We compare X-Trafformer with two categories of baseline models: graph-based models (including ARGCN [46], DCRNN [14], STGCN [25], MSTGCN [24], and ToGCN [47]) and transformer-based models (including Informer [19], Reformer [48], Autoformer [23], and Crossformer [36]):

- *ARGCN* is a relationship extraction model based on attention-guided graph convolutional neural networks. The model obtains new weight matrices through a multi-head self-attention mechanism, each representing a new fully connected graph.
- *DCRNN* leverages diffusion convolutional recurrent neural networks, utilizing bidirectional random walks for spatial correlations and employing an encoder-decoder architecture with predefined sampling for temporal correlations.
- *STGCN* integrates graph convolutional networks and causal convolutional networks to efficiently extract features from both temporal and spatial dimensions, considering causal relationships among time series.
- *MSTGCN* enhances skeleton-based action recognition by introducing multi-scale spatial and temporal graph convolution modules, enabling effective modeling of both short- and long-range dependencies.
- *To-GCN* is a topology optimization-based GCN that leverages latent information by simultaneously optimizing network topology and learning parameters of the fully connected network (FCN).
- *Transformer* is a state-of-the-art neural network architecture renowned for capturing contextual relationships and dependencies in sequential data through self-attention mechanisms, widely employed in natural language processing tasks.
- *Autoformer* excels in long-term time series forecasting, introducing a novel decomposition architecture with an efficient autocorrelation mechanism, achieving state-of-the-art accuracy across diverse practical applications.

- *Informer* is an efficient transformer-based model for long-sequence time series forecasting, overcoming challenges with probsparse self-attention mechanism, attention distillation, and generative style decoder.
- *Crossformer* is a transformer-based model for multivariate time series forecasting, adeptly handling cross-dimension dependencies through strategic integration of dimension-segment-wise (DSW) embedding and a two-stage attention (TSA) layer.

7.5.1.4 Implementation Details

For the Telecom dataset, we utilize the standard transformer model in X-Trafformer. In the case of the Tmall dataset, we employ the Autoformer model. For GNN-based models, experiments are uniformly conducted using the Bigscity framework [49], a comprehensive framework designed for traffic scenarios, equipped with numerous models of different categories. The dataset used in this chapter is formatted to the standard input of the framework, allowing for consistent experiments across various models. The framework permits the customization of parameters for each model, and we set the parameters to the default values mentioned in the original work. For transformer-based methods, we conduct experiments with different input sequence lengths and output sequence lengths. The input sequence lengths vary from $\{16, 32, 64, 96, 150\}$, and the output sequence lengths range from $[1, 150]$. The default output sequence length for comparison is set to 96, which we denote as the "Horizon." The model's dimension is set to 512, the number of heads to 8, and the optimization objective loss to MAE.

7.5.2 Performance Comparison

The experimental results (Tables 7.3, 7.4, and 7.5) show that our framework consistently achieves state-of-the-art results on various datasets. X-Trafformer achieves the top position in 15 out of 20 metrics on the Telecom dataset and achieves the first rank in all metrics on the Tmall dataset. Although some graph-based models outperform us in the MAE metric, this is due to differences in loss functions. Overall, our framework's ability to learn overall correlations among objects and incorporate fine-grained attribute-level information contributes to its superior performance. By introducing more informative features and enhancing representation at multiple granularities, our framework stands out as a top-performing solution.

In the Tmall dataset, existing baseline methods encounter poor results when applying product categories as nodes in GNN-based approaches due to the lack of spatial information with physical significance. Therefore, we choose not to include experimental results in this context. The reason behind this outcome is that GNN-based methods primarily emphasize capturing spatial information, followed by temporal information, which differs from transformer-based methods. It is worth

Table 7.3 Performance comparison between X-Trafformer and various baselines on the Telecom-M

Methods	HORIZON 5		HORIZON 10		HORIZON 25		HORIZON 50		HORIZON 96	
	MAE	MSE	MAE	MSE	MAE	MSE	MAE	MSE	MAE	MSE
ARGCN	0.4080	0.8541	0.3877	0.8926	0.4112	0.8231	0.3997	0.9133	0.4316	0.9974
DCRNN	2.0891	22.4823	2.3954	28.0153	3.0271	44.7790	3.4703	58.5393	4.6334	89.1429
STGCN	1.3299	2.3235	1.3468	2.3395	1.3190	2.3058	1.3269	2.3214	1.3447	2.3320
MSTGCN	2.4547	22.8800	2.7141	27.6687	3.1882	38.2591	3.3831	47.4353	5.7479	91.1200
ToGCN	4.2278	18.4215	4.2263	19.3620	4.2507	19.5284	4.2479	22.5129	5.9642	27.7468
Transformer	0.3803	0.4094	0.3764	0.4077	**0.3764**	0.4074	0.3785	0.4146	0.3820	0.4072
Informer	0.3633	0.3947	0.3736	0.4046	0.3804	0.4123	0.3868	0.4195	0.3831	0.4162
Reformer	0.4477	0.4633	0.4408	0.4603	0.4342	0.4539	0.4301	0.4482	0.4352	0.4555
Autoformer	0.8636	0.7237	0.9104	0.7456	1.0396	0.8023	1.0463	0.8061	1.0040	0.7868
Crossformer	0.5179	0.5215	0.5726	0.5539	0.6224	0.5830	0.6252	0.5815	0.6233	0.5874
X-Trafformer	**0.3612**	**0.3925**	**0.3678**	**0.4029**	0.3788	**0.4048**	**0.3710**	**0.4047**	**0.3660**	**0.4032**

Table 7.4 Performance comparison between X-Trafformer and various baselines on the Telecom-T

Methods	HORIZON 5		HORIZON 10		HORIZON 25		HORIZON 50		HORIZON 96	
	MAE	MSE	MAE	MSE	MAE	MSE	MAE	MSE	MAE	MSE
ARGCN	**0.3778**	0.5514	**0.3806**	0.6512	**0.3765**	0.6479	**0.4021**	0.7508	0.4411	0.7274
DCRNN	0.5517	0.7589	0.5402	0.7338	0.6030	0.8389	0.5934	0.8212	0.6321	1.3254
STGCN	0.8517	3.1075	0.9059	3.8951	1.1359	7.3570	1.2602	9.8440	2.0799	30.3256
MSTGCN	0.9377	1.9467	0.8802	1.8958	0.8474	1.9301	1.0674	2.5936	1.0575	3.0052
ToGCN	3.2445	11.0890	3.2518	11.1331	3.2506	11.1087	3.2678	11.1934	3.2780	11.2459
Transformer	0.4527	0.4644	0.4505	0.4644	0.4517	0.4658	0.4437	0.4534	0.4380	0.4483
Informer	0.4329	0.4492	0.4357	0.4525	0.4412	0.4557	0.4445	0.4551	0.4394	0.4427
Reformer	0.4629	0.4773	0.4567	0.4738	0.4512	0.4649	0.4445	0.4542	0.4323	0.4387
Autoformer	0.8215	0.6807	0.8569	0.6925	0.9742	0.7495	0.9428	0.7491	0.8542	0.7047
Crossformer	0.5179	0.5215	0.5726	0.5539	0.6224	0.5830	0.6254	0.5815	0.6274	0.5974
X-Trafformer	0.4257	**0.4475**	0.4235	**0.4457**	0.4270	**0.4475**	0.4236	**0.4400**	**0.4178**	**0.4268**

Table 7.5 Performance comparison between X-Trafformer and various baselines on the Tmall. The best and second-best results are highlighted in boldface and underlined, respectively. A lower MSE or MAE indicates a better prediction

	HORIZON 5		HORIZON 10		HORIZON 25		HORIZON 50		HORIZON 96	
Methods	MAE	MSE	MAE	MSE	MAE	MSE	MAE	MSE	MAE	MSE
Transformer	0.0342	0.0814	0.0320	0.0779	0.0328	0.0777	0.0333	0.0782	0.0343	0.0798
Informer	0.0470	0.1146	0.0471	0.1169	0.0476	0.1160	0.0484	0.1166	0.0493	0.1171
Reformer	0.0379	0.0906	0.0380	0.0903	0.0387	0.0909	0.0388	0.0914	0.0399	0.0924
Autoformer	<u>0.0251</u>	<u>0.0773</u>	<u>0.0249</u>	<u>0.0756</u>	<u>0.0246</u>	<u>0.0717</u>	<u>0.0245</u>	<u>0.0706</u>	<u>0.0255</u>	<u>0.0721</u>
Crossformer	0.0541	0.1466	0.0540	0.1462	0.0545	0.1456	0.0572	0.1503	0.0572	0.1598
X-Trafformer	**0.0227**	**0.0769**	**0.0233**	**0.0748**	**0.0242**	**0.0705**	**0.0240**	**0.0693**	**0.0249**	**0.0706**

noting that even in the absence of physically meaningful information, our model is still able to extract correlations between objects from the time series data of different objects, enabling the effectiveness of the object layer. Experimental results consistently demonstrate that our work outperforms other baseline methods across various metrics.

7.5.3 Ablation Study

To better demonstrate the performance improvement achieved by employing our framework, we conduct ablation experiments. We use the notation "Metric_Module" to distinguish between different experiments, where "Metric" represents MSE and MAE, and "Module" denotes the following:

"raw": Original transformer-based time series model
"A": Transformer model with only the attribute layer added
"S": Transformer model with only the object module added
"C": Complete version of our framework with both attribute and object modules integrated

Figures 7.4 and 7.5 show that almost all other lines are below the two blue solid lines. Among them, the X-Trafformer, which includes both the object and the attribute layers, exhibits the best performance, showcasing significant improvements compared to the original baseline. Analyzing individual modules, we observe that the inclusion of the object layer leads to substantial performance gains, benefiting almost all transformer-based methods significantly. Additionally, the attribute layer has a more pronounced impact on performance, particularly on large datasets like Telecom-M. This observation highlights that in scenarios with numerous variables, the interrelationships among them can be highly complex, and conventional multivariate methods may not fully account for these complexities. However, our attribute layer can effectively address this issue.

7.5.4 Parameter Study

In this part, we discuss the impact of hyperparameter selection, specifically the spatial representation dimension (dim_s) and the attribute representation dimension (dim_a), on the predictive performance of the framework.

The spatial representation dimension pertains to the size of embedding vectors employed to encode spatial information among objects. By varying the spatial representation dimension, we can explore different levels of spatial granularity in the traffic data. Generally, larger spatial representation dimensions can capture more detailed spatial patterns, but they may also require more computational resources. On the other hand, the attribute representation dimension refers to the

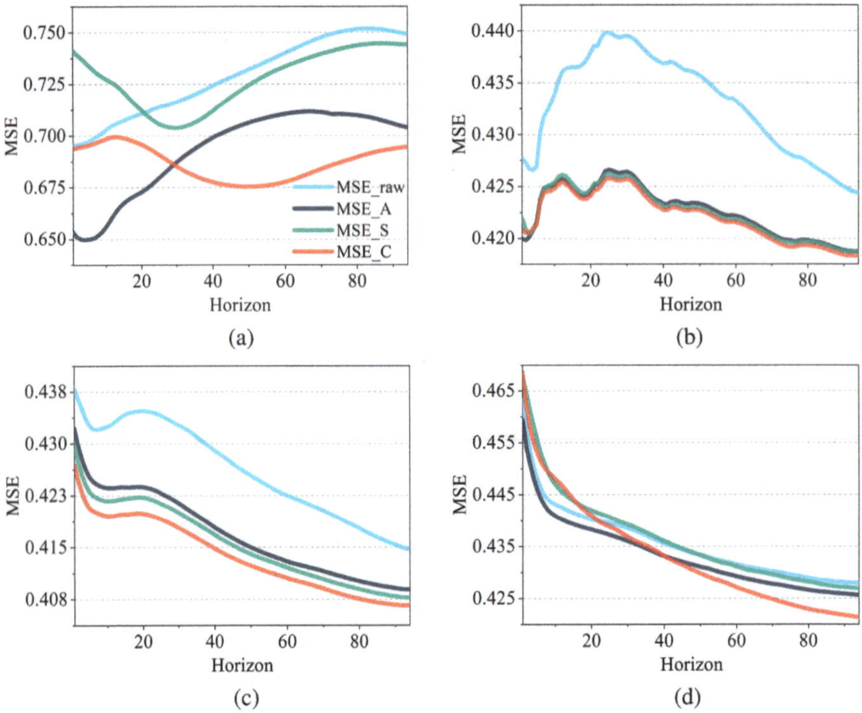

Fig. 7.4 Ablation experiments: Performance comparison of X-Trafformer variants on *Telecom-T*, with mean squared error (MSE) as the evaluation metric. (**a**) Autoformer. (**b**) Informer. (**c**) Transformer. (**d**) Reformer

size of embedding vectors used to represent different variables in the attribute layer. Adjusting the attribute representation dimension allows us to control the information content within the attribute embeddings, thereby influencing the model's ability to consider attribute-related factors in predictions.

In our experiments, while keeping other parameters constant, we systematically vary these hyperparameters and evaluate the framework's performance under different settings. The results, as shown in Tables 7.6 and 7.7, indicate that the choice of spatial representation dimension and attribute representation dimension significantly affects the framework's performance. Specifically, increasing the dimension has the most pronounced impact on performance in the Telecom-M dataset, while its effect on the other two datasets is less evident. Our analysis suggests that the Telecom-M dataset has a higher number of object nodes and larger data dimensions, necessitating higher-dimensional spatial and attribute embeddings for representation. Conversely, the other two datasets have lower dimensions, making it possible to use smaller representation dimensions effectively.

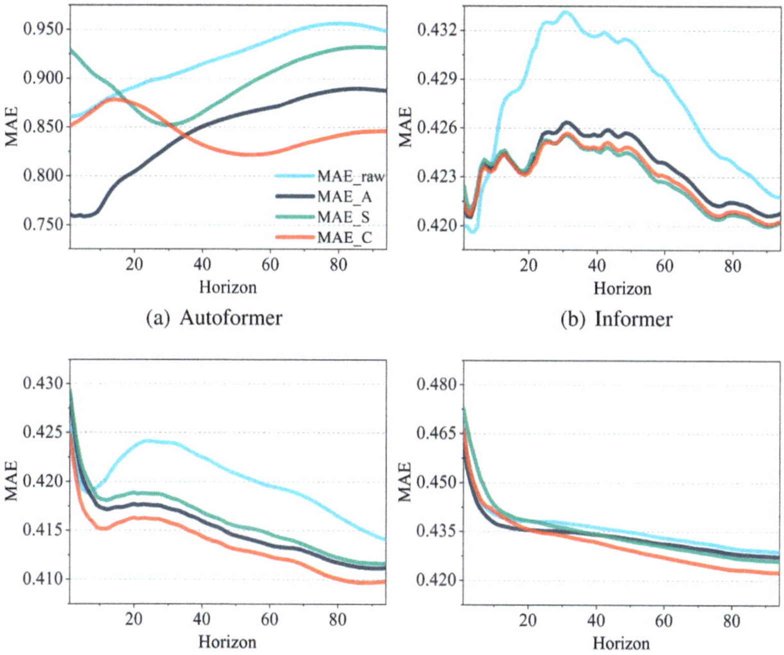

(a) Autoformer (b) Informer (c) Transformer (d) Reformer

Fig. 7.5 Ablation experiments: Performance comparison of X-Trafformer variants on *Telecom-T*, with mean absolute error (MAE) as the evaluation. (**a**) Autoformer. (**b**) Informer. (**c**) Transformer. (**d**) Reformer metric

Table 7.6 Impacts of dim_s on performance

	dim_s	5	10	25	50
Telecom-M	MAE	0.4093	0.3966	0.3831	0.3660
	MSE	0.4110	0.4150	0.4139	0.4032
Telecom-T	MAE	0.4461	0.4318	0.4211	0.4178
	MSE	0.4315	0.4361	0.4287	0.4268
Tmall	MAE	0.0356	0.0356	0.0333	0.0321
	MSE	0.0856	0.0842	0.0789	0.0788

Table 7.7 Impacts of dim_a on performance

	dim_a	8	16	32	64
Telecom-M	MAE	0.4003	0.3864	0.3884	0.3820
	MSE	0.4200	0.4070	0.4000	0.4072
Telecom-T	MAE	0.4214	0.4148	0.4186	0.4176
	MSE	0.4273	0.4202	0.4246	0.4229
Tmall	MAE	0.0350	0.0339	0.0318	0.0316
	MSE	0.0809	0.0791	0.0771	0.0763

Table 7.8 Comparative analysis of computational cost

Dim size in X-Trafformer	Cost (s/ep)	GPU (MB)	MSE	Improve
Raw	31.81	2673	0.4527	–
$dim_s = 10$	36.06	2978	0.4385	3.14%
$dim_s = 50$	37.18	2978	0.4276	5.55%
$dim_a = 16$	53.29	3830	0.4202	7.18%
$dim_a = 64$	64.74	6464	0.4229	6.58%
$dim_s = 10 + dim_a = 64$	75.06	6480	0.4199	7.25%

7.5.5 *Computational Cost*

In this section, we conduct a comprehensive analysis of the model's computational speed and memory consumption for the RQ2 and RQ3 experiments.

From Table 7.8, it is evident that incorporating both the object layer and attribute layer leads to an increase in computational time and memory usage. In-depth analysis reveals that setting the representation dimension of the object layer to 50 results in a 16.8% increase in computational time, but it yields a significant performance improvement of 5.55%. This demonstrates the high cost effectiveness of this approach.

Furthermore, after introducing the attribute layer, the computational time nearly doubles, yet the performance also sees a notable enhancement. Considering that traffic prediction often has fewer real-time demands in practical applications, as it is mostly conducted offline, incorporating the attribute layer becomes a viable choice to maximize performance. Additionally, if memory usage is a concern, using larger dim_s or smaller dim_a representations can achieve substantial performance improvements with only a minor increase in memory consumption.

Overall, these findings demonstrate that the addition of object and attribute layers has an impact on computational efficiency and memory usage. The trade-off between computational cost and performance improvement must be carefully considered, while selecting the appropriate representation dimensions for different layers, to achieve the optimal performance in traffic flow prediction tasks.

7.5.6 *Case Study*

We conduct a case study on the Tmall dataset, focusing on predicting the four attributes (click, collect, cart, and buy) of products with their categories as behavioral objects. Specifically, we select clothing and shoe categories for a prediction horizon of 96. In terms of models, we choose to compare the X-Trafformer using transformer model with the native transformer.

From the experimental results (Fig. 7.6, where Category #0 is clothes and category #1 is shoes.), it can be seen that even when using the native transformer in

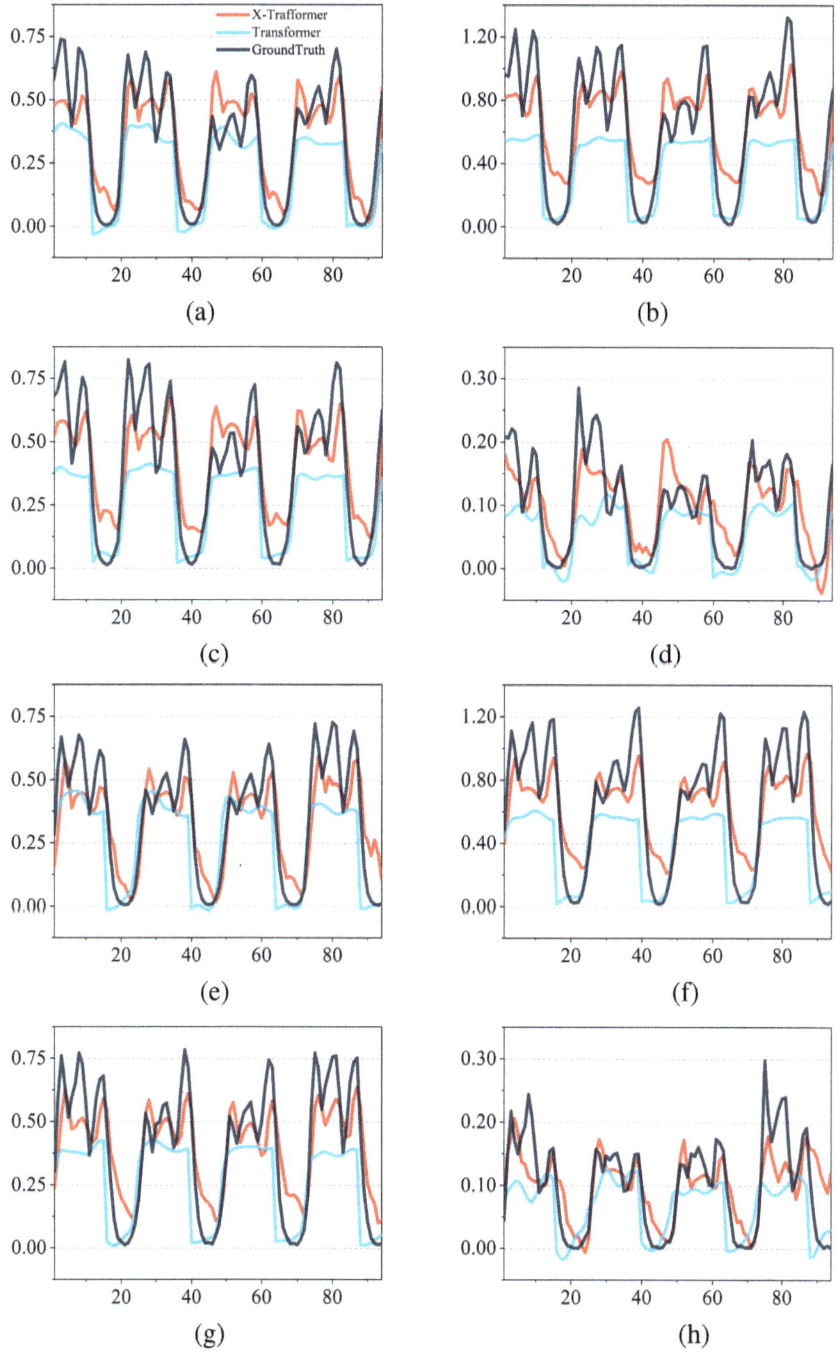

Fig. 7.6 Case Study: Comparisons of prediction results of X-Trafformer, transformer, and ground-truth values on the Tmall data. (**a**) click #0. (**b**) collect #0. (**c**) cart #0. (**d**) buy #0. (**e**) click #1. (**f**) collect #1. (**g**) cart #1. (**h**) buy #1

the behavior layer, X-Traffermer performs much better in prediction accuracy than transformer. X-Trafformer not only captures changing trends effectively but also exhibits finer details compared to the smoother curve of the transformer. In contrast, the transformer's curve appears very smooth, lacking the oscillations and sharpness observed in the X-Trafformer.

Comparing the two sets of graphs, the ground-truth trends are actually quite similar, indicating a high correlation between the categories of clothing and shoes, which aligns with common sense. The transformer model generates notably different predictions because it lacks the capability to learn interactions between objects. In contrast, X-Trafformer, which integrates object modeling, consistently produces similar predictions for analogous objects, leading to a substantial improvement in accuracy.

7.6 Conclusion

This chapter aims to address two extraordinary challenges in traffic prediction for network services, specifically dealing with variable-term traffic series and comprehensive traffic behavior. To tackle these issues, we propose a novel framework named X-Trafformer, which aims to enhance the quality of existing limited traffic data. X-Trafformer conceptualizes traffic as behaviors, thereby transforming the traffic prediction task into predicting behavior events initiated by objects containing specific behavioral attributes. In detail, it employs customized graph structures to capture the spatiotemporal distribution at a coarse-grained object level and the spatiotemporal association at a fine-grained attribute level. Additionally, a behavior layer based on an improved transformer architecture is designed to integrate object and attribute information, enhancing the performance of predicting traffic sequences of arbitrary length. The effectiveness of our framework is demonstrated on real datasets from representative network services. X-Trafformer achieves high prediction performance on limited traffic data with an appropriate computational cost.

Our work achieves the effective utilization of attribute-level and object-level information at a relatively low cost. However, in this process, it inevitably introduces additional computational overhead. In future work, we will focus on optimizing algorithm complexity to reduce the resource requirements of our model. Additionally, we will consider resource allocation and scheduling to achieve lower energy consumption. Particularly in real-world application environments, we will pay attention to cross-platform resource utilization and energy consumption issues.

References

1. K. DeMedeiros, A. Hendawi, M. Alvarez, Sensors **23**(3), 1352 (2023)
2. J. Jiang, C. Han, W.X. Zhao, J. Wang, *Proceedings of the AAAI 2023, Washington, DC, USA, February 7–14* (2023), pp. 4365–4373
3. Percentage of mobile device website traffic worldwide from 1st quarter 2015 to 4th quarter 2022. Statista. https://www.statista.com/statistics/277125/share-of-website-traffic-coming-from-mobile-devices/
4. Y. Yao, B. Gu, Z. Su, M. Guizani, IEEE Trans. Mobile Computing **22**(5), 2837 (2023)
5. Z. Wang, J. Hu, G. Min, Z. Zhao, Z. Chang, Z. Wang, IEEE Trans. Industrial Informatics **19**(4), 5722 (2023)
6. F. Li, Z. Zhang, Y. Zhu, J. Zhang, in *Proceedings of the INFOCOM Workshops 2020, Toronto, ON, Canada, July 6–9, 2020* (2020), pp. 443–448
7. T. Kuber, I. Seskar, N.B. Mandayam, in *Proceedings of the IEEE WCNC 2021, Nanjing, China, March 29–April 1, 2021* (2021), pp. 1–6
8. C. Zhang, S. Dang, B. Shihada, M. Alouini, in *Proceedings of the IEEE INFOCOM 2021, Vancouver, BC, Canada, May 10–13, 2021* (2021), pp. 1–10
9. T.N. Weerasinghe, I.A. Balapuwaduge, F.Y. Li, in *IEEE INFOCOM 2019-IEEE Conference on Computer Communications Workshops (INFOCOM WKSHPS)* (2019), pp. 1–6
10. A. Lazaris, V.K. Prasanna, in *Proceedings of the IM 2019, Washington, DC, USA, April 09–11, 2019* (2019), pp. 19–24
11. Y. Du, H. Huang, Y. Sun, S. Chen, G. Gao, in *Proceedings of the INFOCOM 2021, Vancouver, BC, Canada, May 10–13, 2021* (2021), pp. 1–10
12. Z. Zhang, M. Li, X. Lin, Y. Wang, F. He, arXiv preprint arXiv:1810.10237 (2018)
13. B. Yu, H. Yin, Z. Zhu, in *Proceedings of the IJCAI 2018, July 13–19* (Stockholm, Sweden, 2018), pp. 3634–3640
14. Y. Li, R. Yu, C. Shahabi, Y. Liu, *Proceedings of the ICLR 2018, Vancouver, BC, Canada, April 30–May 3, 2018, Conference Track Proceedings* (2018)
15. W. Jiang, Expert Syst. Appl. **201**, 117163 (2022)
16. G. Lai, W. Chang, Y. Yang, H. Liu, in *Proceedings of the ACM SIGIR 2018, Ann Arbor, MI, USA, July 08–12* (2018), pp. 95–104
17. G. Athanasopoulos, F. Vahid, J. Bus. Econ. Stat. **26**(2), 237 (2008)
18. D. Salinas, V. Flunkert, J. Gasthaus, T. Januschowski, Int. J. Forecast. **36**(3), 1181 (2020)
19. H. Zhou, S. Zhang, J. Peng, S. Zhang, J. Li, H. Xiong, W. Zhang, in *Proceedings of the AAAI 2021, Virtual Event, February 2–9, 2021*, vol. 35 (2021), pp. 11106–11115
20. Y. Yang, J. Lu, Entropy **24**(4), 528 (2022)
21. Z. Wu, S. Pan, G. Long, J. Jiang, X. Chang, C. Zhang, in *Proceedings of the ACM SIGKDD 2020, CA, USA, August 23–27, 2020* (2020), pp. 753–763
22. D. Cao, Y. Wang, J. Duan, C. Zhang, X. Zhu, C. Huang, Y. Tong, B. Xu, J. Bai, J. Tong, Q. Zhang, in *Proceedings of the NeurIPS 2020, December 6–12, 2020, Virtual*
23. H. Wu, J. Xu, J. Wang, M. Long, in *Proceedings of the NeurIPS 2021, December 6–14, 2021, Virtual* (2021), pp. 22419–22430
24. Z. Chen, S. Li, B. Yang, Q. Li, H. Liu, in *Proceedings of the AAAI 2021, Virtual Event, February 2–9, 2021* (2021), pp. 1113–1122
25. H. Han, M. Zhang, M. Hou, F. Zhang, Z. Wang, E. Chen, H. Wang, J. Ma, Q. Liu, in *Proceedings of the IEEE ICDM 2020, Sorrento, Italy, November 17–20, 2020* (2020), pp. 1052–1057
26. N. Ramakrishnan, T. Soni, in *Proceedings of the IEEE ICMLA 2018, Orlando, FL, USA, December 17–20, 2018*, pp. 187–193
27. Z. Tian, IEEE Trans. Intell. Transp. Syst. **22**(9), 5566 (2021)
28. Q. Wu, K. He, X. Chen, S. Yu, J. Zhang, IEEE/ACM Trans. Networking **30**(3), 1255 (2022)
29. F.A. Gers, J. Schmidhuber, F. Cummins, Neural Comput. **12**(10), 2451 (2000)

30. R. Vinayakumar, K.P. Soman, P. Poornachandran, in *Proceedings of the ICACCI 2017, Udupi (Near Mangalore), India, September 13–16, 2017* (2017), pp. 2353–2358
31. C. Zhang, H. Zhang, D. Yuan, M. Zhang, IEEE Commun. Lett. **22**(8), 1656 (2018)
32. T.N. Kipf, M. Welling, in *Proceedings of the ICLR 2017, Toulon, France, April 24–26, 2017, Conference Track Proceedings* (2017)
33. J. Wang, J. Tang, Z. Xu, Y. Wang, G. Xue, X. Zhang, D. Yang, in *Proceedings of the IEEE INFOCOM 2017, Atlanta, GA, USA, May 1–4, 2017* (2017), pp. 1–9
34. Q. Liu, J. Li, Z. Lu, IEEE Commun. Lett. **25**(10), 3325 (2021)
35. A. Vaswani, N. Shazeer, N. Parmar, J. Uszkoreit, L. Jones, A.N. Gomez, L. Kaiser, I. Polosukhin, in *Proceedings of the NeurIPS 2017, December 4–9, 2017, Long Beach, CA, USA* (2017), pp. 5998–6008
36. Y. Zhang, J. Yan, in *Proceedings of the ICLR 2023, Kigali, Rwanda, May 1–5, 2023* (2023)
37. Z. Zhou, M. Shojafar, M. Alazab, F. Li, IEEE Trans. Industr. Inform. **18**(12), 8967 (2022)
38. Y. Cui, K. Cao, J. Zhou, T. Wei, IEEE Trans. Comput. Aided Des. Integr. Circuits Syst. **42**(5), 1518 (2023)
39. Z. Zhou, M. Shojafar, M. Alazab, J.H. Abawajy, F. Li, IEEE Trans. Green Commun. Networking **5**(2), 658 (2021)
40. J. Tang, M. Qu, M. Wang, M. Zhang, J. Yan, Q. Mei, in *Proceedings of the WWW 2015, Florence, Italy, May 18–22, 2015* (2015), pp. 1067–1077
41. B. Perozzi, R. Al-Rfou, S. Skiena, in *Proceedings of the ACM SIGKDD 2014, New York, NY, USA—August 24–27, 2014* (2014), pp. 701–710
42. T. Yang, L. Hu, C. Shi, H. Ji, X. Li, L. Nie, ACM Trans. Inf. Syst. **39**(3), 32:1 (2021)
43. L. Hu, T. Yang, C. Shi, H. Ji, X. Li, in *Proceedings of the EMNLP-IJCNLP 2019, Hong Kong, China, November 3–7, 2019* (Association for Computational Linguistics, New York, 2019), pp. 4820–4829
44. T. Italia, *Telecommunications—SMS, Call, Internet—MI* (2015). https://doi.org/10.7910/DVN/EGZHFV
45. A. Group, *Rec-Tmall: Brand and Item Data from Tmall Recommendation Contest* (2022). https://tianchi.aliyun.com/dataset/140281
46. L. Bai, L. Yao, C. Li, X. Wang, C. Wang, in *Proceedings of the NeurIPS 2020, NeurIPS 2020, December 6–12, 2020, virtual* (2020)
47. L. Yang, Z. Kang, X. Cao, D. Jin, B. Yang, Y. Guo, in *Proceedings of the IJCAI 2019, Macao, China, August 10–16, 2019* (2019), pp. 4054–4061
48. N. Kitaev, L. Kaiser, A. Levskaya, in *Proceedings of the ICLR 2020, Addis Ababa, Ethiopia, April 26–30, 2020* (2020)
49. J. Wang, J. Jiang, W. Jiang, C. Li, W.X. Zhao, in *Proceedings of the 29th International Conference on Advances in Geographic Information Systems* (New York, NY, USA, 2021), pp. 145–148

Chapter 8
Detecting Evolving Behavior in Open World

Abstract The convenience offered by the Internet accelerates the evolution of fraudulent behavior during facilitating the rapid development of online payment services. Fraudsters can change their behavior patterns frequently and at a low cost in the online space, allowing them to evade regulatory oversight. This poses a significant challenge for meticulously trained learning-based security applications for fraud detection and can lead to serious social security risks. To stay ahead of the rapidly evolving fraud, researchers have been exploring models with low latency and fast response capabilities to effectively combat fraudulent behavior. Unfortunately, the evolving fraud is not only reflected in the drift of their superimposed risk features but also in the openness of their category. The interweaving of open category and concept drift accelerates the process of existing security methods becoming powerless. This book proposes EvoFD, an online evolving fraud detection framework to enable continual learning to cope with undercurrent surges of evolving fraud. The core idea of EvoFD is to weaken the bias caused by the *anchoring effect* on the learned information. It learns in an online streaming fashion by using instructive representations as anchors. The advantages of our method are validated by extensive experiments over the real-world dataset from a prestigious bank.

8.1 Introduction to Online Behavior Evolving

With the explosive growth of web services, electronic commerce has gained in popularity. People tend to consume and trade through online payment services [1]. Online payment services bring convenience to people's lives. Meanwhile, the convenience of the Internet accelerates the evolution of fraudulent behavior. Fraudsters can easily exploit the anonymity, global reach, and high-speed transmission in online space to engage in fraudulent behavior more easily. Moreover, they can leverage various technological means to change their behavior patterns frequently and at a low cost. Various cybercrimes involving online payment frauds have caused enormous losses to both online payment platforms and customers [2, 3]. To safeguard the interests of normal users and eliminate social security risks, a robust and efficient fraud detection solution is critical to the development of online payment services [4].

C. Wang, *Computational Structural Behavior*,
https://doi.org/10.1007/978-981-95-1013-9_8

Behavior-based fraud detection is a promising paradigm in the anti-fraud scenario [5, 6]. Particularly, the online spaces record a wealth of information about user payment behavior, when compared to offline spaces. This information includes detailed records of various actions taken during the payment process, which provides a breeding ground for the implementation of efficient security solutions based on behavior. Deep learning techniques have been employed by researchers to extract complex features from massive behavioral data, and various deep learning methods have been proposed to address fraud detection problems [3, 7, 8]. Unfortunately, most of these learning-based applications with superior performance depend on a strong assumption of independent and identically distributed between training and test samples (static learning paradigm). With the assumption of independent and identical distribution, the model is trained over a static training dataset and then gets frozen and deployed for inference without making changes to the model parameters again once the training is completed.

This hinders their applicability in real-world scenarios, where fraud patterns are complex and even evolving [9]. The distribution of the offline dataset is usually inconsistent with that of online data. Therefore, this strict assumption is usually not valid in open world, where transactions are typically displayed in a streaming fashion and behavior patterns evolve over time [10, 11]. Malicious patterns may even switch abruptly and significantly in adversarial environments against security systems. Failure to detect and adapt to such shifts can result in a high number of false positives and false negatives, thereby requiring fraud detection models to have new capabilities to combat evolving fraud [12].

Sensing the evolution of fraud is expensive because it faces not only the drift of superimposed risk features but also the openness of fraud category. The interweaving of them renders any existing technologies powerless.

Inspired by biological systems, continual learning is used to break the strong barrier between the static training and testing phase [13]. It allows the model to be continually optimized on new data (e.g., unseen fraud categories) instead of being trained on a static dataset [14]. The goal is to develop models that do not stop learning but keep updating the model parameters over time [15], just as humans perceive the world with a tendency to learn concepts sequentially. Due to the catastrophic forgetting of artificial neural networks, one of the greatest bottlenecks of continual learning is to retain inherent knowledge and absorb new knowledge, i.e., accumulating knowledge over time [16].

Recent works about continual learning focus on addressing the issue of catastrophic forgetting with task sequence learning [17, 18]. There are three classes of methods, i.e., replay methods, regularization-based methods, and parameter isolation methods [19]. Most of these researches about continual learning have opted for a specific experimental setup, i.e., a sequence of distinct tasks is given [20]. Based on the setup, continual learning makes significant progress where data can be shuffled within a task to guarantee i.i.d. condition, and the model can observe the data multiple passes to enhance model learning in the way of *offline learning*. However, the development of them in this specific setup depends on the perception of task boundaries. The changes in data distributions, including not only *open category* but also *concept drift*, might suddenly happen without the indication of task identifier. In real-life online payment scenarios, data needs to be processed in a streaming fashion [21]. Therefore, existing continual learning methods cannot handle well with online fraud detection.

This book proposes an online evolving fraud detection framework for online payment services, named EvoFD, to weaken the bias caused by the anchoring effect on the learned information. The framework adopts the online setup for model learning to cope with the transaction streams in the online payment services. In the framework, the training and evaluation can be conducted simultaneously or independently. The model has the ability of continual learning and can adapt to task-free transaction streams. In this setup, the distribution could experience a gradual or sudden change, and task identifiers and boundaries are not provided. The online evolving fraud detection framework has three devised parts, i.e., replay memory, class anchor, and class-balancing loss. In the replay memory, the framework stores a subset of observed transactions, which will be replayed when new transaction instances are trained. It shares the same importance to each class, so all observed classes of transactions are allocated the consistent memory capacity. The framework employs reservoir sampling to obtain a uniform distribution from the entire data stream. The balanced replay memory facilitates the consolidation of old knowledge and model learning in transaction streams. To represent the characteristics of changing classes, the framework maintains the class anchors to approximate the centers of mass in the latent space. To avoid the obsolescence of class anchors as the representation network changes, the momentum-based update is conducted to optimize the class anchors. The representation network embeds the features and class anchors into a unified normalized space. The nearest neighbor-based classification infers the transaction classes with the normalized cosine similarity. In addition, for the imbalanced transaction streams, it devises a class-balancing loss, named smooth weighted cross-entropy (swce) loss. The swce loss adaptively adjusts the class weights with the distributions of the transaction streams evolving but avoids the frequent jitters of class weights. Several experiments are deployed in collaboration with a prestigious bank in China. The results show that EvoFD outperforms representative baselines in detecting fraudulent behaviors in the evolving fraud scenario.

8.2 From Continual Learning to Fraud Detection

8.2.1 Continual Learning

With the recent increased interest in neural networks, continual learning has received more attention [22, 23]. Continual learning studies the learning from an infinite data stream with the goal of gradually accumulating knowledge and utilizing it for future learning [24]. Continual learning is generally categorized into three families, i.e., replay methods, regularization-based methods, and parameter isolation methods [19].

Replay methods store samples in raw format or generate virtual (pseudo) samples with a generative model [25]. These previous samples are replayed when learning new samples to alleviate forgetting. The class incremental learner iCaRL [18] stored a subset of exemplars which are the best approximate class centers in the latent feature space. CLEAR [26] leveraged off-policy learning and behavioral cloning from replay to enhance stability and on-policy learning to preserve plasticity. GEM [25] aimed to constrain new task updates to not interfere with previous tasks by projecting the estimated gradient direction on the feasible region outlined by previous task gradients through first-order Taylor series approximation. Atkinson et al. [27] used a generative adversarial network to generate items so that the deep network can learn to classify sequentially with the representation of previous tasks. Shin et al. [28] proposed a framework with a cooperative dual model architecture consisting of a deep generative model and a task-solving model so that training data for previous tasks can easily be sampled and replayed in a new task.

Regularization-based methods introduce an additional regularization term in the loss function to consolidate previous knowledge when learning over new data. Silver and Mercer [20] proposed to use the previous task model to output given new task input images for improving new task performance. LwF [17] used the previous model outputs as soft labels for previous tasks to mitigate forgetting and transfer knowledge. Rannen et al. [29] built an under-complete autoencoder for each task to preserve previous knowledge and control the features in their low-dimension submanifold. EWC [30] selectively slowed down learning on the weights important for old tasks to remember old knowledge. VCL [31] developed a general framework for continual learning that fused online variational inference (VI) in Monte Carlo VI for neural networks so that it can successfully train both deep discriminative models and deep generative models in complex continual learning settings. MAS [32] computed the importance of the parameters of a neural network in an unsupervised and online manner to allow increased flexibility and online user adaptation.

Parameter isolation methods preserve different model parameters for each task to prevent any possible forgetting. Rusu et al. [33] proposed to leverage prior knowledge via lateral connections to previously learned features to avoid forgetting. Lee et al. [34] proposed to employ an autoencoder to infer tasks and apply parameter isolation to prevent catastrophic interference. RCL [35] devised a continual learning method by searching for the best neural architecture for each coming task via sophisticatedly designed reinforcement learning strategies. Expert Gate [36] learned tasks/experts sequentially and introduced a set of gating autoencoders to learn a representation for the task at hand, and at test time to forward the test sample automatically to the relevant expert. PackNet [37] exploited redundancies in large deep networks to free up parameters that can then be employed to learn new tasks and conducted iterative pruning and network retraining to sequentially pack multiple tasks into a single network. Serrà et al. [38] proposed a task-based hard attention mechanism that preserved previous tasks' information without affecting the current task's learning.

8.2.2 Fraud Detection

Fraud detection has been investigated by many researchers for a long time. Early works on fraud detection mainly focus on rule-based approaches [39]. Rosset et al. [40] presented a special-purpose rule discovery system, a two-stage system based on the adaption of the rule generator, with an additional rule selection mechanism. Brause et al. [41] generated and weight the association rules to detect credit card fraud.

Fraud patterns become more and more complicated over time [42]. This paradigm is not only expensive, and the effectiveness of rules induced by the expert system might deteriorate. Recently, researchers have turned to data-driven fraud detection methods with the development of big data and machine learning [43, 44]. Supervised learning-based methods have been widely applied for fraud detection [45, 46]. Yuan et al. [47] combined deep neural networks and spectral graph analysis. Utilizing topology information, they developed a deep autoencoder and convolutional neural network to detect fraud. Zhu et al. [45] proposed a hierarchical explainable network to model users' behavior sequences to improve the performance of fraud detection. Cheng et al. [48] devised a spatial-temporal attention-based graph network. They learned the temporal- and location-based transaction graph features by a graph neural network and employed the spatial-temporal attention and detection network to detect card fraud. Xi et al. [46] captured the hierarchical information with a two-level structure over the user's event sequence. SAFE [49] adopted a recurrent neural network to handle user activity sequences and mapped dynamic user activities to survival probabilities to detect fraud. Liu et al. [44] devised a tree-like structure to reorganize the user behavioral data and a neural method to utilize the behavioral tree for fraud transaction detection.

There are also many works focusing on semi-supervised and unsupervised fraud detection technologies [50, 51]. Carcillo et al. [52] proposed a taxonomy of streaming active learning strategies and conducted an extensive comparison of these techniques for fraud detection accuracies. Carcillo et al. [53] proposed a hybrid method that combined supervised and unsupervised methods to improve fraud detection accuracy. Kim et al. [54] adopted a novel hierarchical cluster based on deep neural networks to detect more detailed fraud and fraud of whole data. ScoreGAN [55] devised a semi-supervised GAN method that made use of both review text and review rating scores in the generation and detection process, where scores were incorporated through information gain maximization into the loss function.

8.3 Background

8.3.1 Scenario

Online payment service providers face significant threats from malicious fraud, such as telecom fraud, phishing websites, and coupon trimming within criminal groups. Existing machine learning methods have made considerable progress in detecting fraudulent online payment transactions. Unfortunately, the change in behavioral patterns poses one of the most frustrating challenges for these methods, which are built on the closed-world assumption that the same distribution exists between training and test samples. Over time, normal users may alter their behavioral patterns, and the booming electronic economy has accelerated the evolution of normal user payment habits. It renders the knowledge previously learned by machine learning models outdated, leading to significant performance degradation if this concept shift is not detected and adjusted for. In real-world deployments, failure to detect and adapt to this concept shift results in a high number of misreports and omissions. Additionally, in the anti-fraud scenario, concept drift in malicious fraudulent behavior is also common, as malicious patterns may suddenly and significantly switch in adversarial environments over time. For example, the introduction of new patches, anti-fraud technologies, and transaction protocols can alter the malicious behavior patterns of fraudsters. Moreover, fraudsters actively and frequently change their strategies to evade fraud detection solutions.

Consider the following fraud behaviors in online payment services. In transaction activities, fraudsters often use various means to transfer funds from victims' accounts for illegal profit. These fraud means emerge endlessly and change rapidly. The characteristics of fraud behaviors are varied with the types and means of fraud. The illustration of this evolving fraud scenario is presented in Fig. 8.1. One day, the fraudsters take a means (fraud A) to defraud others' assets. After a variety of complaints and reports from users, the online payment platform notices this type of fraud. To avoid greater losses, the platform trains a data-driven model over this kind of collected fraud transactions and normal transactions. This fraud

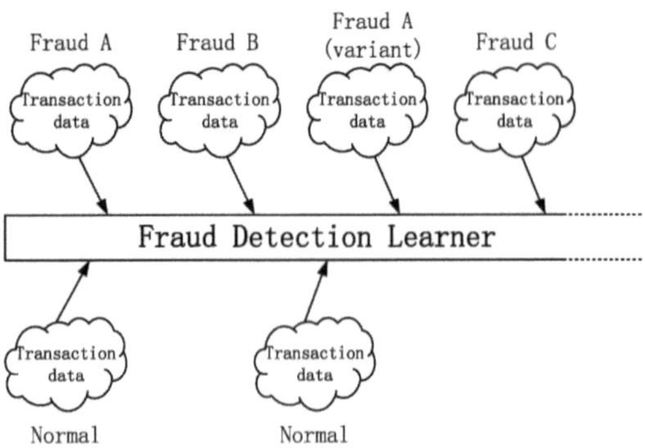

Fig. 8.1 The illustration of evolving fraud scenario

detection model, which can identify fraud A, is deployed in the online environment to resist fraudsters. Then the behaviors of the fraudsters are intercepted by the fraud detection system due to the identifying of this type of fraudulent characteristics. The fraudsters abandon previous means and discover a new type of trick (fraud B) to evade detection. After the accumulation of late-model fraud, the online payment platform gradually finds this type of scam. It retrains the fraud detection model for the new fraud pattern over the new transactions and historical transactions including previous fraud and normal data. With the ineffectiveness of fraud B, the fraudsters turn to upgrade the previous means to develop a variant fraud (fraud A-variant). The online payment platform tends to retrain the detection model for new patterns after collecting sufficient fraudulent transactions. However, the fraudsters will continue to develop new fraud. The games between fraudsters and the fraud detection system are always progressing.

8.3.2 Challenge

Machine learning models have exhibited superior performance on individual tasks, such as image classification and object recognition. However, the impressive results are obtained static learning paradigm, and these models are incapable of adapting to the dynamic environment over time. In online payment services, transactions are generated in the form of a data stream, which requires the fraud detection system to adapt continuously and keep learning over time. For neural networks, a major challenge is catastrophic forgetting. The performance on previously learned tasks or domains would significantly degrade over time as new tasks or domains are learned. This directly results in a general problem, i.e., the stability-plasticity dilemma, where stability means retaining previous knowledge, but plasticity refers to the ability to accept new knowledge. The online fraud detection in online payment services is a delay-sensitive task. The latency of fraud prediction will affect the prompt interception of fraudulent transactions.

While capturing fraud in online payment services, the fraud detector would have to recognize not only fraudulent but also benign transactions. In the real world, normal transactions are much more than fraudulent transactions. This produces another important problem, i.e., imbalance learning. The models would learn the prior information of the class proportion in the training set, so that the actual prediction would have a bias toward the majority samples. Especially in online payment scenarios, normal transactions occupy absolute predominance in the transaction streams. The imbalance problem aggravates the learning difficulty. Moreover, this difficulty is further complicated by the dynamically changing imbalanced distribution.

8.3.3 Problem Analysis

Typical fraud detection methods are based on static learning and i.i.d. assumptions. As illustrated in Fig. 8.2, for the given task T, the learner is trained over the training examples (X, Y) from an unknown probability distribution D. Here, X is the vectors of input space, and Y is the corresponding class labels indicating the types of transactions, such as the normal or the fraudulent. Under the assumption that the testing data X' follows the same distribution, the learner can give the prediction labels of X'. In this paradigm, typical fraud detection methods train the detector over the historical transaction records extracted from the database. Then the detector is deployed in the online environment. When ongoing transactions are taken in the system, the detector can infer the types of transactions for risk decisions. The transactions are accumulating over time, and their distributions might undergo changes. While sufficient transactions are collected, the detector will be retrained over the new data.

Continual learning is widely researched to deal with learning in sequences. Generally, task-based continual learning methods assume that data comes in tasks, with task boundaries identified. Then the training procedure can be divided into consecutive phases. The learner updates its meta-knowledge in the training procedure and tries to avoid catastrophic forgetting. The learner is typically evaluated after the stream has been observed entirely. During the evaluation procedure, the learner requires or infers the task identifiers that indicate which task the instances to be predicted belong to [25]. The task identifiers can simplify the continual learning significantly. The task refers to an isolated training phase with a batch of data that belongs to a new group of output space, domain, or class [56]. Consequently, this class of continual learning scenarios can be categorized into three kinds, i.e., task-incremental learning (Task-IL), domain-incremental learning (Domain-IL), class-incremental learning (Class-IL) [57]. The task-based continual learning is represented as a task sequence, $\{T_1, T_2, ..., T_t, ...\}$. We use $T_t = \{\mathscr{X}^{(t)}, \mathscr{Y}^{(t)}, t\}$ to represent a task, where $\mathscr{X}^{(t)} = \{x_i^{(t)}\}_{1 \leq i \leq n}$ is a set of learning samples and $\mathscr{Y}^{(t)} =$

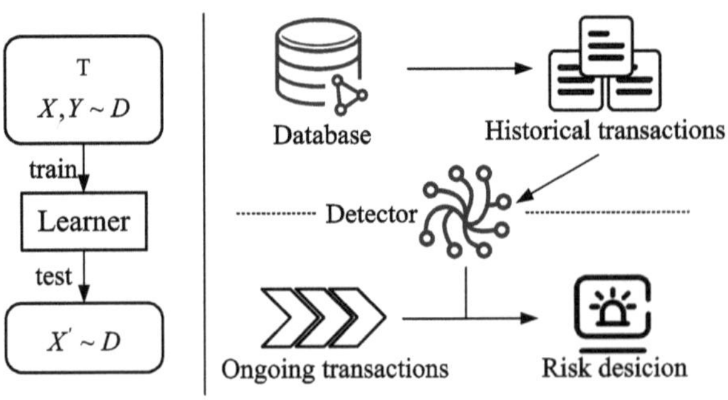

Fig. 8.2 Typical fraud detection setup

Table 8.1 The task-based continual learning scenarios

Scenario	Task ID	Output space	Marginal probability
Task-IL	Explicit	$\{\mathscr{Y}^{(t)}\} \neq \{\mathscr{Y}^{(t+1)}\}$	$P(\mathscr{Y}^{(t)}) \neq P(\mathscr{Y}^{(t+1)})$
Domain-IL	No	$\{\mathscr{Y}^{(t)}\} = \{\mathscr{Y}^{(t+1)}\}$	$P(\mathscr{Y}^{(t)}) = P(\mathscr{Y}^{(t+1)})$
Class-IL	Implicit	$\{\mathscr{Y}^{(t)}\} \subset \{\mathscr{Y}^{(t+1)}\}$	$P(\mathscr{Y}^{(t)}) \neq P(\mathscr{Y}^{(t+1)})$

$\{y_i^{(t)}\}_{1 \leq i \leq n}$ is the set of output targets. The marginal probability distribution of input space between the old task and the new task is generally changing, specifically $P(\mathscr{X}^{(t)}) \neq P(\mathscr{X}^{(t+1)})$. The main differences of continual learning scenarios come from the output space, which are summarized in Table 8.1. In Task-IL, the task identifiers are always explicitly provided, although the distributions of input space and output space are changing. In Domain-IL, without the identifiers of tasks, the learning difficulty is simplified by the unchanged distributions of output space. In Class-IL, the entire classes are divided into disjoint subsets so the task identifiers can be inferred. These setups simplify the actual continual learning scenarios with some priori information. However, this task-based setup is hard to adapt to data streams and is rarely encountered in practice applications.

8.4 Anchor-Based Behavior Evolving Framework

EvoFD is an online evolving fraud detection method in online payment scenarios. The pipeline of the online evolving fraud detection framework is presented in Fig. 8.3. At each time step t, the transaction stream is drawn from the transaction records of all accounts at this moment, where the data $\{\mathscr{X}_t, \mathscr{Y}_t\}$ obeys the distribution $\mathscr{D}_t$. The detector learns continually with the transaction stream progressing and the data distribution evolving. Given an account of a user, its transactions can be formalized as a sequence in the timeline, i.e., $R^u = \{r_1^u, r_2^u, ..., r_t^u\}$. All transactions at each time step constitute the data stream. We consider a lengthy (even

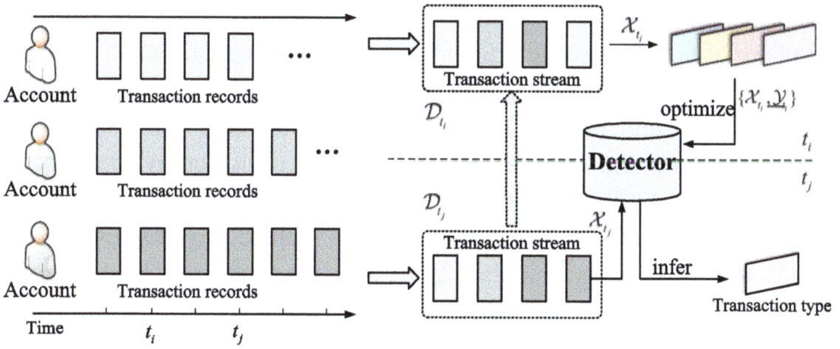

Fig. 8.3 The pipeline of online evolving fraud detection

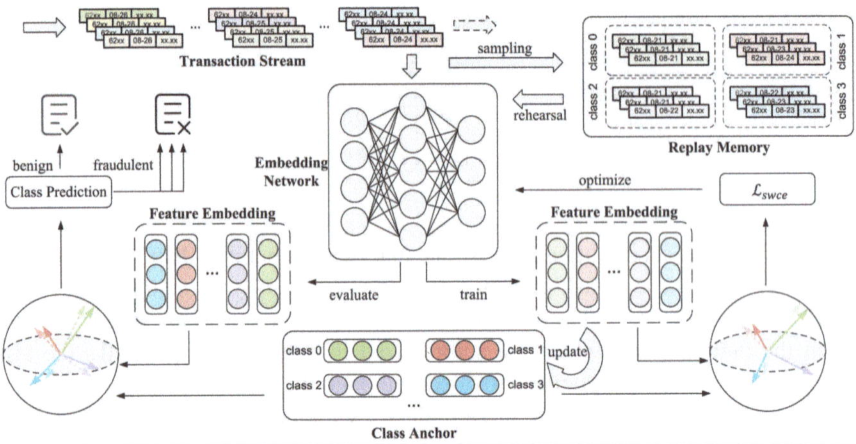

Fig. 8.4 Illustration of our approach for online fraud detection

potentially infinite) nonstationary transaction stream $\mathscr{S}$. At each time step t, the system receives a new set of transactions $\{r_t^1, r_t^2, \ldots, r_t^u\}$. After feature processing and labeling, the learner obtains a batch of samples along with labels $\{\mathscr{X}_t, \mathscr{Y}_t\}$ drawn in a non-i.i.d. (independently and identically distributed) manner from a distribution $\mathscr{D}_t$ that could experience a gradual or sudden change and switch to distribution $\mathscr{D}_{t+1}$ [58]. When inferring, an ongoing transaction is processed and sent to the learner for inference. The labeled transactions can be utilized to optimize the learner continually. The learner might need additional resources after observing the incoming data, such as the replay memory to store samples. The training and inferring of the learner can be simultaneous and independent.

Our goal is to learn a model f parameterized by θ that minimizes the prediction errors on the ongoing transaction samples. It aims to update and accumulate knowledge continuously. Given a replay memory $\mathscr{M}$ of finite size, the learner at each time step reduces the empirical risk based on the newly received small batches of samples and the stored samples in $\mathscr{M}$. Our methods comprise three main components: replay memory, class anchor, and class-balancing loss. The workflow of our proposed approach is illustrated in Fig. 8.4. Our system observes a continuous transaction stream. At a time step, the system receives a small batch of transactions. The rehearsal is conducted to train the learner with the transaction samples in the replay memory and the incoming batch. The devised class-balancing loss is employed to optimize the embedding network. Then the replay memory and class anchors are updated by the incoming transactions. When deploying the model, the transactions are judged by the similarity between feature and class embeddings. The transactions will be assigned the class label with the most similar anchor.

8.4.1 Replay Memory

We design a replay-based method in continual learning as this kind of method shows better performance when faced with long sequences [15]. The rehearsal methods store a subset of previous instances and explicitly retrain on the limited stored samples and new samples [19]. For generalization, we do not make strong assumptions about the distributions of data streams. The data streams might be highly imbalanced, especially in online payment services, where the amount of fraudulent samples is much smaller than the normal samples.

Here, we assume equal importance on the replay memory of each class. We divided the replay memory equally to all classes, i.e., $M/|\mathscr{C}|$ per class, where M is the capacity of replay memory and $\mathscr{C}$ is the set of observed classes. This scheme allocates consistent memory capacity for all observed classes, making the memory management independent of the characteristics of the data stream and alleviating the impact of the potentially imbalanced distribution. We randomly draw a subset of the data stream of each class to approximate its latent distribution and update the replay memory with the date stream evolving. The scheme of updating replay memory is shown in Algorithm 6. We dynamically adjust the capacity of each class buffer to fully utilize the whole capacity of the memory and keep it balanced. When reducing the capacity of a class from m' to m, we will keep the first m samples and discard the others. When the class buffer is full, we will employ a reservoir sampling [59] to replace the existing instances. As the storage carrier of old knowledge, replay memory aims to approximate the distribution of the observed stream at as less cost as possible. The sampling can be conducted in an infinite data stream and ensure equal probabilities on all observed instances. The scheme is more adaptable to online scenarios when compared with the herding technique [18]. When receiving a new incoming set of instances, we conduct a rehearsal based on the received batch and the samples retrieved randomly from the replay memory.

8.4.2 Class Anchor

We train an embedding network f parameterized by θ and maintain class anchors $\mathscr{P}$ for every class, where f_θ aims to project the feature vectors of instances to a latent space. The class anchor represents each class by approximating the cluster center of the corresponding class in the latent space. We enforce the class anchors to L^2-normalized vector to eliminate the bias caused by the difference in magnitudes. iCaRL [18] stores the exemplars and computes the mean of their features extracted by the embedding network. It would be computationally expensive in the online setup because the embedding network always changes, and the class prototypes need to be recomputed at any time. The instances in the evaluation are predicted by the

Algorithm 6: Updating replay memory

Input: replay memory $\mathscr{M}$, memory capacity M, incoming batch B_t, observed classes $\mathscr{C}$, observed sample count N

foreach $(x_i, y_i) \in B_t$ **do**
  if $y_i \notin \mathscr{C}$ **then**
    % Initialize new class replay memory $\mathscr{M}^{y_i} \leftarrow \{\}$
    $\mathscr{C} \leftarrow \mathscr{C} \cup \{y_i\}$, $N^{y_i} \leftarrow 0$
    $m \leftarrow M/|\mathscr{C}|$
    foreach $\mathscr{M}^c \in \mathscr{M}$ **do**
      $\mathscr{M}^c \leftarrow \{x_k | x_k \in \mathscr{M}^c, k = 1, 2, ..., m\}$
    end
  end
  $N^{y_i} \leftarrow N^{y_i} + 1$
  if $|\mathscr{M}^{y_i}| < m$ **then**
    $\mathscr{M}^{y_i} \leftarrow \mathscr{M}^{y_i} \cup \{x_i\}$
  end
  else
    $k \leftarrow \text{Uniform}(1, N^{y_i})$
    if $k \leq |\mathscr{M}^{y_i}|$ **then**
      $\mathscr{M}_k^{y_i} \leftarrow x_i$
    end
  end
end

nearest neighbor-based classification. We assign the feature vector x_i of instance the class label with the most similar anchor:

$$y^* = \arg\min_{c \in \mathscr{C}} ||f_\theta(x_i) - p_c||, \tag{8.1}$$

where $\mathscr{C}$ is the set of observed classes and $p_c \in \mathscr{P}$ is the anchor vector of class c.

We consider updating the class anchor based on the incoming batch B_t and replayed exemplars $B_{\mathscr{M}_t}$. The data stream shifts rapidly in the nonstationary environment, so we finally compute the class anchors based on the momentum update:

$$p_c \leftarrow mp_c + (1-m)\frac{1}{|\mathscr{X}_t^c|}\sum_{x \in \mathscr{X}_t^c} f_\theta(x), \tag{8.2}$$

where $m \in [0, 1]$ is the momentum, and $\mathscr{X}_t^c$ is the union of B_t^c and $B^c_{\mathscr{M}_t}$, i.e., the instances of class c. After the update, we renormalize the class anchors to ensure the unit magnitude. When encountering a new class, we generate a random L^2-normalized vector as the class anchor. The class anchor will progressively be updated to approximate the center of actual distribution and reflect the shift of the class concept.

8.4.3 Class-Balancing Loss

In a typical classification network, the predicted probability of an instance x is computed by the softmax activation function as follows:

$$P_i(x) = \frac{\exp(w_i^\top f_\theta(x))}{\sum_j \exp(w_j^\top f_\theta(x))}, \tag{8.3}$$

where feature extraction part f_θ and weight vectors of last layer $w_1, w_2, ..., w_{|\mathscr{C}|} \in \mathbb{R}^d$ constitute the classification network. The weight vectors can be regarded as class embeddings. We could straightforwardly use the outputs as final probabilities, but the network needs to be dynamically adjusted as new class occurs in the online scenarios. To avoid the cost of model expansion, we employ only the feature extraction part to project the instances to a latent space of fixed dimensions. We also L^2-normalize the embedding vectors, i.e., $\bar{v} = v/||v||_2$, to make them comparable to class anchors. When the vectors are normalized to unit length, the squared euclidean distance can be obtained from the cosine distance:

$$||v_1 - v_2||^2 = 2 - 2\langle v_1, v_2\rangle. \tag{8.4}$$

Here, we measure the similarity by the cosine distance. Given the class anchors $\mathscr{P}$, the predicted probability is computed as follows:

$$P_i(x) = \frac{\exp(\lambda\langle p_i, \bar{f}_\theta(x)\rangle)}{\sum_{c\in\mathscr{C}} \exp(\lambda\langle p_c, \bar{f}_\theta(x)\rangle)}, \tag{8.5}$$

where hyperparameter λ controls the probability distribution over classes and $\langle v_1, v_2\rangle = v_1^\top v_2$ measures the cosine similarity between two normalized vectors.

For a multi-classification task, we could adopt the standard cross-entropy loss:

$$\mathscr{L}_{ce} = -\frac{1}{|\mathscr{B}|} \sum_{(x_i,y_i)\in\mathscr{B}} \Big[\sum_{c\in\mathscr{C}} \delta_{c=y_i} \log P_c(x_i) + \delta_{c\neq y_i} \log(1 - P_c(x_i))\Big], \tag{8.6}$$

where $\mathscr{B} = B_t \cup B_{\mathscr{M}_t}$ is the training batch, and δ is the indicator function. The standard cross-entropy loss is generally used in the balanced classification task. In online payment services, normal transactions are much more than fraudulent transactions. The transaction data stream is severely imbalanced. Under this optimization objective, the model would be biased toward the majority class, i.e., normal transactions, which would significantly reduce the detection rate of fraud. For the imbalanced data stream, we propose the smooth weighted cross-entropy (swce) loss. Firstly, we compute the smooth class weight with the clip operation:

$$s_i = \begin{cases} \log(\mu \frac{|B|}{|B^i|}) & \log(\mu \frac{|B|}{|B^i|}) > 1 \\ 1 & otherwise \end{cases}, \tag{8.7}$$

where the smooth factor μ smoothes the class weights and $B^i \in B$ is subset corresponding class i. The small μ reduces the sensitivity toward class imbalance, which avoids the frequent jitters of class weights. The smooth class weights are varied adaptively with the distribution of data stream changes. Then, we compute the class-balancing loss on batch B:

$$\begin{aligned} \mathcal{L}_B = -\frac{1}{|B|} \sum_{(x_i, y_i) \in B} s_{y_i} \Big[\sum_{c \in \mathcal{C}} \delta_{c=y_i} \log P_c(x_i) \\ + \delta_{c \neq y_i} \log (1 - P_c(x_i)) \Big]. \end{aligned} \tag{8.8}$$

Finally, the objective function can be obtained:

$$\mathcal{L}_{swce} = \beta \mathcal{L}_{B_t} + \mathcal{L}_{B_{\mathcal{M}_t}}, \tag{8.9}$$

where β is the loss balance weight to adjust the inclination between previous and new knowledge. The process of optimizing the online fraud detector is described in Algorithm 7.

8.5 Evaluation of EvoFD

In this part, we perform a variety of experiments to investigate the proposed method for fraud detection in a real-life online payment scenario. We describe the experimental setup, including dataset, implementation, and metrics in Sect. 8.5.1. Through the empirical evaluation, we mainly aim to answer the following three research questions:

RQ1: To what extent can our method surpass existing methods when dealing with real-world online fraud that evolves over time? (Sect. 8.5.2)

RQ2: How much contribution can each component in our method bring? (Sect. 8.5.3)

RQ3: What impact will the key parameters in our method have on the overall performance? (Sect. 8.5.4)

Algorithm 7: Optimizing online fraud detector

Input: class anchors $\mathscr{P}$, replay memory $\mathscr{M}$, observed classes $\mathscr{C}$

foreach $B_t = \{(x_1, y_1), ..., (x_{|B_t|}, y_{|B_t|})\} \in S$ **do**
  foreach $(x_i, y_i) \in B_t$ **do**
    if $y_i \notin \mathscr{C}$ **then**
      foreach $j \in (1, 2, ..., d)$ **do**
        $p_{y_i}[j] \leftarrow \text{Uniform}(0, 1)$
      end
      $\mathscr{P} \leftarrow \mathscr{P} \cup \{\bar{p}_{y_i}\}$
    end
  end
  Sampling $B_{\mathscr{M}_t}$ from replay memory $\mathscr{M}$
  $\mathscr{B} \leftarrow B_t \cup B_{\mathscr{M}_t}$
  Compute $\mathscr{L}_{swce}$ using Eq. (8.9)
  $\theta \leftarrow \theta + \alpha \nabla \mathscr{L}_{swce}$
  foreach $c \in \mathscr{C}_{\mathscr{B}}$ **do**
    Update class anchor p_c using Eq. (8.2)
  end
  Update replay memory using Algorithm 6
end

8.5.1 Experimental Setup

8.5.1.1 Dataset Description

The experiments are conducted on a real-world online payment dataset from a prestigious commercial bank. This dataset was collected solely for scientific research purposes in June 2017, and users were informed that their data would be used for research after being anonymized. Meanwhile, in compliance with data protection regulations, the data provider employed de-identification and anonymization techniques on sensitive data. These techniques encompassed, but were not limited to, data encryption and data hashing. As a result, the original data became unavailable, and no sensitive privacy information was utilized in the experiment.

In the dataset, each online payment transaction is comprised of 64 features. These features can be categorized into two types. (1) User features, which are related to the personal information provided by users to ensure account security and prevent unauthorized transactions. In this dataset, there are 17 features associated with user personal information, including user ID, account ID, bank card type, and mobile phone number. (2) Behavior features, which encompass the payment operations recorded by the bank's payment platform during online transactions. These include device information (such as operating system, MAC address, and IP address), transaction attributes (transaction amount and time), and transaction object information (merchant ID), among others.

Due to reasons such as user refusal to grant authorization, many features are considered missing and cannot be collected. Table 8.2 provides an analysis of

Table 8.2 Statistical information on missing values in the features

Missing rate	Counting value	Missing rate	Counting value
>0.2	36	>0.6	28
>0.3	32	>0.7	16
>0.4	32	>0.8	10
>0.5	30	>0.9	8

Table 8.3 Features of transactions

Field	Quantity	Description
Account ID	74, 105	The card number of transaction, the unique identification of account
Transaction time	804, 354	The time when the transaction happened, beginning with year, accurate to second
Transaction amount	125, 375	The amount of money in the transaction
Transaction type	11	The type of the transaction
Account balance	20	The balance of the account before the transaction happened
Place number	327	The issuing place of banking card
Frequent IP	2	The status indicating whether the current IP is commonly used by the account
Check	15	The verification tool at the process of the transaction

the proportion of missing values for each feature. It is evident that nearly half of the features have a missing rate exceeding 40%, with 10 features even having a missing rate exceeding 80%. Moreover, the overall missing rate of features across the entire dataset is approximately 40.6%. This high rate of missing features results in sparse data with limited available and extractable information, thereby presenting significant challenges for training anti-fraud models.

To address this issue, we employ a random forest classifier to evaluate the importance of features and filter out irrelevant fields. As a result, we primarily select eight key features for our experiment, as summarized in Table 8.3. Among the collected features, there are 35 numerical features of float type, while the remaining features are character-based. Specifically, we include two numerical features, transaction amount and account balance, as well as six character-based features. Here, Account ID represents the bank card number used during the payment process and typically consists of a unique string. In our dataset, there are over 70,000 different card numbers recorded. Transaction time is recorded with second-level precision, resulting in over 800,000 distinct timestamps indicating when the transactions occurred. This implies that multiple transactions can occur simultaneously within the same second.

Online payment service providers face serious threats from malicious fraud, such as telecommunications fraud, phishing website scams, and coupon clipping within organized gangs. In this chapter, the data provider labeled the transactions through phone surveys, complaints, and retrospective rules. Each online transaction was classified as either normal or specific fraud types, with fraud transactions primarily

Table 8.4 Types of labels

Label	Description
0	Normal transaction by the system judgment and secondary confirmation
1	Other kinds of fraud
2	Phishing website
3	Zombie account
4	Telecommunication fraud
5	Trojan virus

categorized as phishing website fraud, zombie account fraud, telecommunications fraud, fraud caused by Trojan viruses, and other types. The label descriptions are presented in Table 8.4.

User behavior patterns may undergo sudden changes over time, particularly in hostile environments where malicious fraud patterns are more likely to switch significantly against fraud detection systems. However, the outstanding performance of machine learning-based solutions is built upon the closed-world assumption of independent and identically distributed (i.i.d.) data between training and testing samples. To simulate real-world online fraud detection scenarios, we divide the transaction sequences into five subsequences in chronological order. It is important to note that our method does not require task boundaries, but we partition the stages for comparison with other methods. The transaction data stream consists of five consecutive stages, with each stage described by a subsequence of transactions. The model predicts the transaction type after each time step update. Performance evaluation is deployed at each stage. In each stage, data samples are randomly split into a training set (70%) and a testing set (30%), with validation samples randomly selected from the training set to determine the optimal parameters.

The statistical data of the samples at each stage is illustrated in Fig. 8.5. The stage sequences have typically data-incremental characteristics in the real world. The transactions of class 4 appear from Stage 1 to Stage 2, and the transactions of class 5 appear from Stage 2 to Stage 3. From Stage 3 to Stage 5, the transactions of the observed classes continuously accumulate.

8.5.1.2 Implementation Details

All models are implemented with PyTorch and trained on Tesla V100 GPUs. We adopt an MLP with two hidden layers of 128 units as the embedding network. The dimension of the embedding vector is set to 16. The ReLU activation function is applied behind each hidden layer. The networks are trained by SGD with a batch size 10. The grid search is conducted to search for the optimal setup.

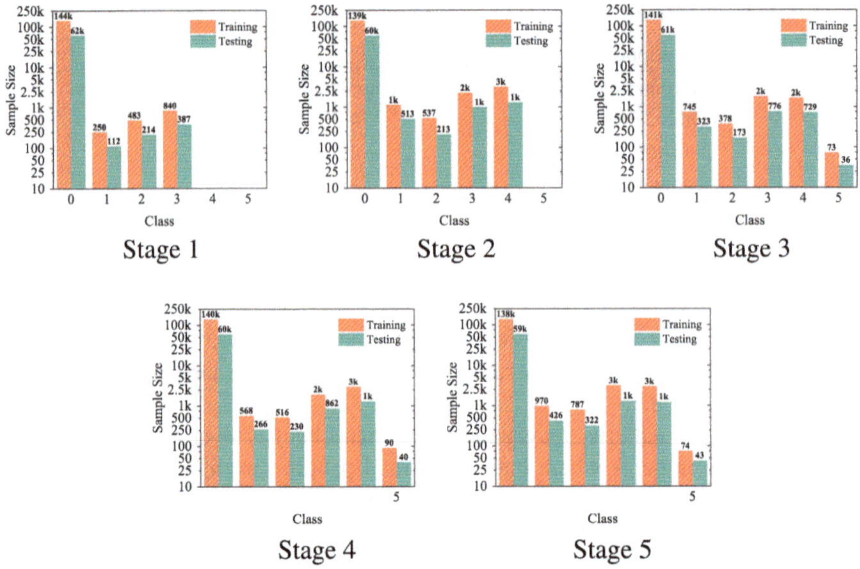

Fig. 8.5 Sample sizes of stages

8.5.1.3 Evaluation Metrics

In general, accuracy is used to evaluate the performance of the multi-classification task. However, in online payment services, the number of normal transactions is significantly higher than the number of fraudulent transactions. In the case of imbalance problems, accuracy is biased toward the majority class and may not reflect the real performance of the model. Therefore, we use the F1 score to evaluate the performance of each class. The overall performance evaluation is done using the macroF1 score, which assigns equal importance to each class and is calculated as the arithmetic mean of the F1 score for each class. Meanwhile, we further adopt the Mathew correlation coefficient (MCC) to evaluate our work.

8.5.2 Comparison with Baselines

To evaluate the performance of our proposed methods, the following methods are chosen as baselines:

- *Fine-tuning* trains the model continuously upon the new tasks without any forgetting avoidance strategy.
- *Joint-online* jointly optimizes the model by interleaving samples from each task. The learner shuffles the entire samples to follow the i.i.d. setting and conducts a single-pass optimization to follow the online setting.

- *Joint-offline*, similar to the Joint-online model, operates in an offline setting. The learner undergoes multiple passes through the entire dataset during training. It is widely acknowledged as the *upper bound* in the context of continual learning.
- *Reservoir* is a replay baseline, which employs the reservoir sampling to manage the replay memory.
- *GEM* replays memory in a constraint optimization, which has been shown to be a strong baseline in the online setting.
- *SurpriseNet* tackles the challenge of catastrophic interference by employing parameter isolation methods and leveraging autoencoders to learn cross-task knowledge.

The comparison results are shown in Tables 8.5 and 8.6. We conduct five runs of experiments for every method by setting different random seeds, where the random seeds affect data sampling, model training, and so on. We report the F1 score on each class and the macroF1 on all classes with standard deviation.

In general, we can find the difficulty of learning online in the gap in performance between Joint-online and Joint-offline. It can be observed that the gap widens with the progress of stage, from about 19% at Stage 1, to about 38% at Stage 3, to about 42% at Stage 5. In some situations, compared with other methods, Joint-online generally presents the poorest performance, especially 0% of F1 score on class 5 when encountering the new class. It can be proven how difficult the online learning setup is to model learning. In addition, for the overall performance on each stage, fine-tuning also gives worse results in most of the stages, confirming that the catastrophic forgetting is indeed preventing the model from improving in the continual learning. Our method (EvoFD) achieves superior performance in general, outperforming other baselines except for the theoretical upper bound, Joint-offline, at the transfer of stages. In the continual optimization, our method closes the gap with Joint-offline performance from 22% at Stage 1, to 9% at Stage 3, to 11% at Stage 5. Meanwhile, compared to the other baselines, our method shows the lowest standard deviation most of the time, indicating that our proposed method has stable performance in online fraud detection.

Specifically, from Stage 1 to Stage 2, our method shows better performance in the newly observed class, achieving 85% of F1 score. The performance is much better than other baseline methods, almost reaching the result of Joint-offline. Our method also outperforms most of the other baselines on the old classes. From the perspective of macroF1, EvoFD is in a backward state at Stage 1 and achieves the leading performance at Stage 2. This shows that in the case of class increment, our method can well balance the knowledge between the newly observed class and the old classes that have been learned.

From Stage 2 to Stage 3, our method also performs better on the newly encountered class than other baseline methods. The overall performance is still at the leading level although the performance in the old classes has a slight decline. From Stage 1 to Stage 3, it can be found that there are new categories between the transitions of each stage. In addition to consolidating the inherent performance of the old categories, the model also needs to show strong plasticity. Our method can

Table 8.5 The performance between our method and baselines at Stage 1–3

Stage	Method	F1 score (%)							
		Class 0	Class 1	Class 2	Class 3	Class 4	Class 5	MacroF1 (%)	MCC (%)
No. 1	Fine-tuning	99.49 ± 0.21	38.39 ± 8.13	69.89 ± 3.52	26.31 ± 2.27	–	–	58.52 ± 2.67	51.46 ± 6.43
	Joint-online	99.58 ± 0.08	31.13 ± 6.52	57.83 ± 15.71	27.42 ± 3.89	–	–	53.99 ± 4.76	47.84 ± 2.29
	Joint-offline	99.69 ± 0.01	64.29 ± 4.41	82.94 ± 1.70	44.87 ± 1.57	–	–	72.95 ± 0.80	65.76 ± 0.46
	Reservoir	99.57 ± 0.08	45.56 ± 4.09	72.65 ± 2.41	26.69 ± 4.00	–	–	61.12 ± 1.32	54.04 ± 2.43
	GEM	99.60 ± 0.13	42.67 ± 4.62	69.39 ± 4.41	27.09 ± 2.04	–	–	59.69 ± 2.77	52.53 ± 3.03
	SurpriseNet	99.60 ± 0.04	33.02 ± 14.39	69.38 ± 8.54	25.95 ± 2.15	–	–	56.99 ± 5.81	52.54 ± 4.73
	EvoFD	98.02 ± 0.18	25.26 ± 1.65	63.81 ± 4.78	17.65 ± 0.75	–	–	51.19 ± 1.38	46.11 ± 1.73
No. 2	Fine-tuning	99.50 ± 0.04	21.35 ± 9.64	21.01 ± 10.80	39.06 ± 11.33	65.22 ± 4.15	–	49.23 ± 3.51	72.05 ± 1.69
	Joint-online	99.48 ± 0.12	8.84 ± 5.04	17.13 ± 16.38	30.09 ± 9.55	60.90 ± 5.00	–	43.29 ± 5.44	69.84 ± 2.48
	Joint-offline	99.63 ± 0.01	67.61 ± 5.83	71.60 ± 3.04	69.71 ± 1.69	85.90 ± 1.55	–	78.89 ± 2.25	86.42 ± 1.24
	Reservoir	99.49 ± 0.04	36.06 ± 6.29	42.28 ± 8.28	46.89 ± 4.37	69.09 ± 1.67	–	58.76 ± 2.56	75.34 ± 1.55
	GEM	99.48 ± 0.01	25.45 ± 5.97	23.52 ± 9.05	33.06 ± 14.60	60.01 ± 8.13	–	48.30 ± 1.54	71.69 ± 0.61
	SurpriseNet	99.33 ± 0.05	53.89 ± 19.24	10.20 ± 7.06	49.84 ± 8.26	79.95 ± 5.14	–	58.28 ± 6.37	75.09 ± 3.20
	EvoFD	98.63 ± 0.03	64.33 ± 4.39	46.96 ± 6.69	46.68 ± 2.17	85.30 ± 1.01	–	68.38 ± 0.25	79.73 ± 0.09
No. 3	Fine-tuning	99.22 ± 0.27	32.19 ± 7.92	12.96 ± 9.33	33.03 ± 5.61	65.95 ± 3.15	9.70 ± 9.30	42.17 ± 2.06	66.35 ± 1.71
	Joint-online	99.41 ± 0.17	4.12 ± 2.75	10.64 ± 12.24	22.86 ± 8.07	59.26 ± 3.22	0.00 ± 0.00	32.72 ± 3.22	61.63 ± 1.69
	Joint-offline	99.59 ± 0.02	66.30 ± 2.22	55.45 ± 3.01	57.51 ± 2.18	82.63 ± 1.13	62.26 ± 3.28	70.62 ± 1.13	80.08 ± 0.66
	Reservoir	99.46 ± 0.06	43.74 ± 12.52	24.14 ± 7.97	40.94 ± 4.46	69.40 ± 2.34	16.87 ± 12.26	49.09 ± 4.00	69.79 ± 2.76
	GEM	99.44 ± 0.00	46.98 ± 0.83	22.13 ± 5.00	34.34 ± 6.34	69.85 ± 1.82	24.44 ± 19.04	49.53 ± 5.50	70.10 ± 3.72
	SurpriseNet	99.19 ± 0.96	2.70 ± 0.93	24.43 ± 12.50	21.83 ± 10.68	48.72 ± 8.49	22.15 ± 12.67	36.35 ± 6.45	55.70 ± 3.44
	EvoFD	98.96 ± 0.13	61.91 ± 2.07	43.88 ± 0.40	38.76 ± 2.15	78.14 ± 2.41	48.13 ± 2.56	61.63 ± 0.51	76.60 ± 0.29

Table 8.6 The performance between our method and baselines at Stage 4–5

Stage	Method	F1 score (%)							
		Class 0	Class 1	Class 2	Class 3	Class 4	Class 5	MacroF1 (%)	MCC (%)
No. 4	Fine-tuning	99.38 ± 0.12	34.96 ± 3.84	13.51 ± 6.36	37.96 ± 2.18	62.51 ± 10.65	57.49 ± 18.76	50.97 ± 4.39	77.12 ± 1.47
	Joint-online	99.39 ± 0.20	17.64 ± 4.09	11.50 ± 9.08	19.46 ± 6.00	70.63 ± 2.23	0.00 ± 0.00	36.44 ± 2.01	70.55 ± 1.52
	Joint-offline	99.61 ± 0.01	62.34 ± 3.80	63.06 ± 3.51	63.74 ± 1.41	88.80 ± 1.35	87.72 ± 2.50	77.54 ± 0.23	84.58 ± 0.09
	Reservoir	99.47 ± 0.99	35.69 ± 6.97	31.50 ± 13.22	42.73 ± 4.63	75.62 ± 3.45	62.90 ± 13.12	57.98 ± 3.84	78.40 ± 2.15
	GEM	99.26 ± 0.21	32.12 ± 3.39	23.77 ± 8.87	35.72 ± 5.79	65.01 ± 0.31	31.23 ± 17.89	47.85 ± 4.84	75.57 ± 2.12
	SurpriseNet	97.24 ± 3.20	0.14 ± 0.32	8.57 ± 4.36	2.38 ± 3.73	23.97 ± 20.09	5.18 ± 3.80	22.91 ± 5.26	46.00 ± 12.65
	EvoFD	99.15 ± 0.10	49.50 ± 5.19	47.45 ± 6.04	49.03 ± 2.97	81.45 ± 0.65	53.55 ± 6.73	63.36 ± 2.12	79.35 ± 1.02
No. 5	Fine-tuning	99.40 ± 0.10	26.10 ± 6.67	31.28 ± 5.40	51.33 ± 7.10	62.93 ± 9.33	44.04 ± 11.73	52.51 ± 1.98	77.38 ± 1.32
	Joint-online	99.37 ± 0.24	13.91 ± 7.54	9.23 ± 6.18	28.73 ± 9.14	57.77 ± 1.91	0.00 ± 0.00	34.83 ± 1.48	67.68 ± 1.16
	Joint-offline	99.65 ± 0.02	66.37 ± 2.99	71.99 ± 2.22	78.67 ± 0.53	85.41 ± 1.40	61.62 ± 6.93	77.29 ± 1.63	88.23 ± 1.19
	Reservoir	99.42 ± 0.19	41.13 ± 6.03	40.02 ± 8.25	60.04 ± 6.14	71.84 ± 2.94	39.19 ± 12.38	58.61 ± 3.93	79.46 ± 1.48
	GEM	99.50 ± 0.04	42.66 ± 7.51	45.31 ± 2.20	62.10 ± 2.80	72.06 ± 1.69	54.24 ± 0.00	62.65 ± 1.81	80.18 ± 0.92
	SurpriseNet	97.21 ± 2.29	3.35 ± 5.17	7.32 ± 7.82	29.07 ± 6.80	22.20 ± 4.83	2.60 ± 1.77	26.96 ± 1.90	41.09 ± 9.43
	EvoFD	99.36 ± 0.06	54.52 ± 3.83	56.89 ± 3.68	65.35 ± 0.86	75.67 ± 1.76	45.97 ± 3.25	66.29 ± 1.30	81.41 ± 0.73

achieve a good balance in such class-incremental scenarios so as to achieve higher overall results.

From Stage 3 to Stage 5, it can be observed that the data of every category is constantly accumulating, and the model needs to continually update the category cognition of the model with new data, and it is expected to use the previous cognition of categories to assist the learning of new knowledge. Our method achieves a better improvement in most of the classes compared to other baselines. We can observe that although the performance of fine-tuning in some classes is decreasing, that of our method is still increasing. It implies that our method can accumulate knowledge and utilize the old knowledge to enhance performance. SurpriseNet shows a significant performance decline in Stage 3 to Stage 5, due to its extremely low performance on individual class, which lowered MacroF1. Because Joint-offline learns the whole data distribution from a global perspective, the inconsistent performance of Joint-offline at each stage can also reflect the different distribution of data at different stages. The Joint-offline serves as a baseline model trained on all historical data in an offline setting. The reason why EvoFD's performance is lower than the Joint-offline model can be attributed to the inherent nature of stream-based learning. Unlike the Joint-offline model, EvoFD is designed to handle dynamic and evolving data streams. It adapts to new instances and updates its knowledge incrementally over time. This adaptability sometimes comes at the cost of slightly lower performance compared to the Joint-offline model, which benefits from training on the complete historical data. However, it is important to note that EvoFD's primary strength lies in its ability to continuously learn and adapt to changing fraud patterns in real time. It provides a more practical solution for fraud detection in dynamic environments.

8.5.3 Ablation Study

To further investigate the working mechanism of our proposed approach, we perform experiments to study the effect of individual aspects of the method. We alter the loss function of our method by using Eq. (8.6). EvoFD-CE adopts the vanilla cross entropy to compute the loss. For comparison, the results of fine-tuning are presented as a baseline.

Figure 8.6 shows the confusion matrices of different methods at the end of stages. For visual presentation, the values in confusion matrices are normalized to [0, 1] by row. It can provide further insight into the effect of each component. The diagonal entries present the correct predictions of methods, and the off-diagonal entries present the incorrect predictions. Fine-tuning tends to identify samples as newly learned classes due to the forgetting of the old knowledge. EvoFD and EvoFD-CE perform better on the diagonal entries. EvoFD-CE has a bias toward normal class and gives worse results on correct predictions. The results imply that the ability to accumulate knowledge and keep on learning is necessary for fraud detector in

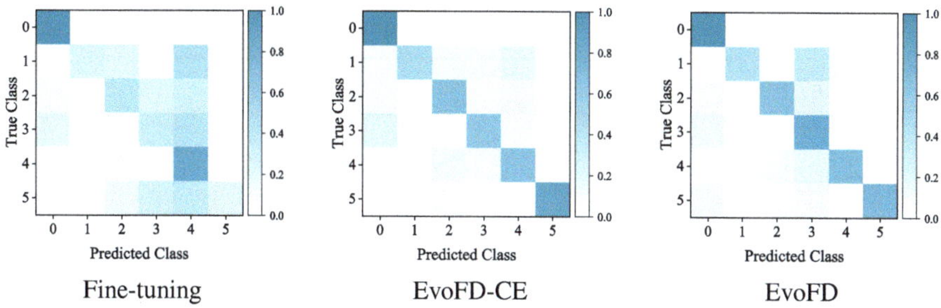

Fig. 8.6 The confusion matrices of different methods at the end of stages. The values are normalized by row

the transaction stream. And our class-balancing loss can balance the trends toward majority classes and minority classes.

Figure 8.7 summarizes the results of the macroF1 score on each stage. The comparison of EvoFD and EvoFD-CE shows that the class-balancing loss is advantageous in most situations. In general, EvoFD and EvoFD-CE present similar trends. In Stage 1, there are four classes, and the imbalance is mild, where EvoFD-CE is slightly better on the performance. In the cold start phase, there are insufficient samples in the replay memory, which might cause the performance disadvantage compared to fine-tuning. In Stage 2, the performance of fine-tuning degrades because of catastrophic forgetting and class imbalance. Our method can remember the previous knowledge to facilitate the learning of new knowledge so that they exhibit the superior performance. Then with the emergence of new classes, the standard cross-entropy loss tends to the majority class, and the gap in performance between EvoFD-CE and EvoFD widens.

8.5.4 Differential Analysis

In this part, we adjust the parameters to study how the setup of each part affects the model.

Memory Size The setup of replay memory capacity is significant in online payment fraud detection systems. High-capacity replay memory means the system stores more transaction samples, which brings about more storage load and security risk. The size of replay samples would affect the model learning. We set several memory sizes, i.e., 100, 200, 300, 500, and 1000, to present the results. Figure 8.8a illustrates the effect of different sizes of replay memory. We present the macroF1 score of the detector in each stage. From 100 to 300, the model benefits from the larger memory size, showing that more exemplars can approximate better the distribution of historical data to help the model retain more old knowledge. But

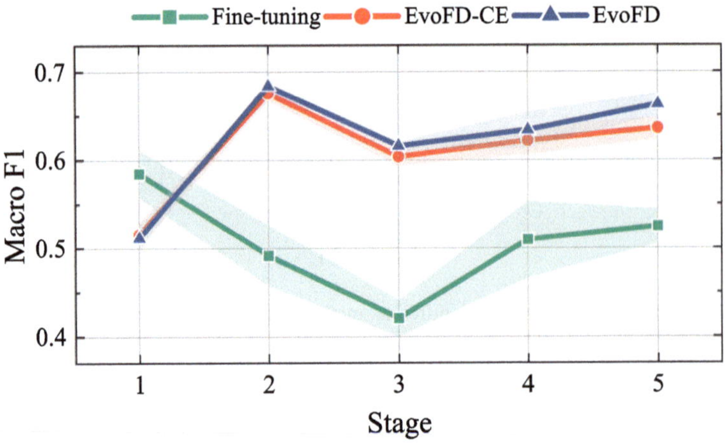

Fig. 8.7 Ablation study

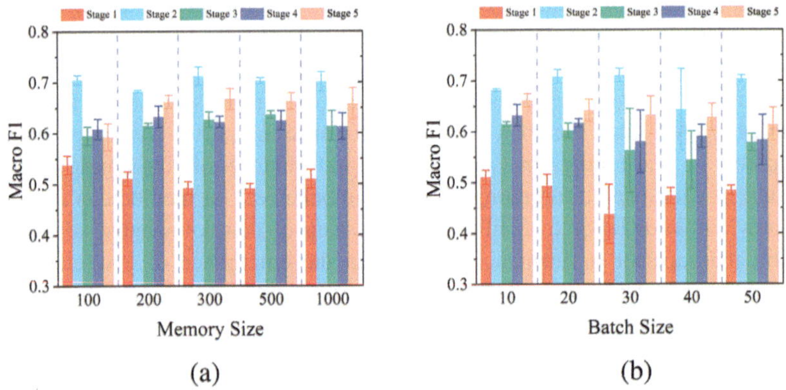

Fig. 8.8 The performance of EvoFD over different parameters

when memory size is more than 300, the samples of minority class are insufficient, and more old samples bring more interferences to the learning of new knowledge.

Batch Size In online learning scenarios, the batch size is generally set to a tiny value to simulate learning from data streams. The setup of the larger batch size conforms more to offline learning scenarios. We aim to investigate how performance is affected by batch size. We choose several batch sizes, i.e., 10, 20, 30, 40, 50. The setup of large batch size is mostly used in offline learning. Figure 8.8b presents the effect of different batch sizes. It can be observed that with the increase in batch size, the performance decreases. The performance for small batches is more robust, which might result from the more frequent update of anchors and the network.

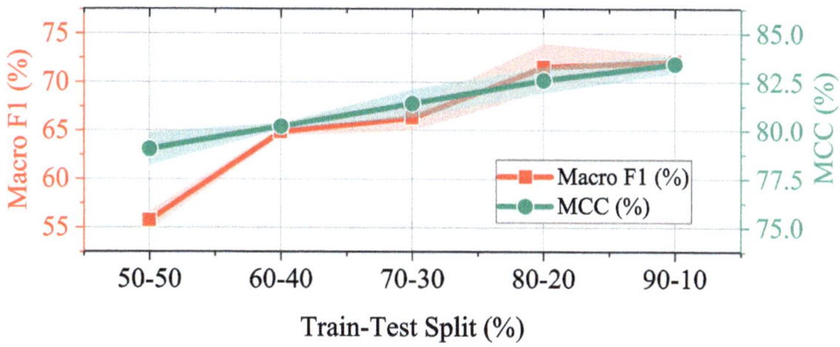

Fig. 8.9 The performance of EvoFD under varying train-test splits

Furthermore, we adjust the train-test split ratios to gain valuable insights into the impact of data allocation between the training and testing sets on the model's performance.

Train-test Split Increasing the proportion of training set samples allows for more samples to be used for model training, resulting in a significant increase in the number of samples for extremely low proportions of fraudulent classes. The varying quantities of different fraudulent samples can affect model learning. In our study, we focus on examining the performance of EvoFD in Stage 5. We adopt the following train-test split ratios: 50-50, 60-40, 70-30, 80-20, and 90-10. Figure 8.9 presents the detection performance of our model on Stage 5 under different train-test splits. We observe that as the proportion of the training set increases, both the macroF1 score and MCC show an overall upward trend. Particularly, in the 80-20 and 90-10 splits, we can observe high macroF1 scores (exceeding 0.7) and MCC (exceeding 82.5), indicating better results for the model under these splits. Additionally, we note that the 50-50 split exhibits a lower macroF1 score. This may be due to the smaller training set, which is insufficient to provide enough fraudulent samples to train a model with higher generalization ability. Therefore, under these splits, the model performs poorly when faced with a small number of fraudulent sample classes. Furthermore, we observe that the macroF1 score shows greater fluctuation, while MCC increases uniformly. In imbalanced data, the model may exhibit lower accuracy and recall when detecting smaller classes, thus reducing the macroF1 score. Therefore, different class sample distributions under different train-test splits may have a significant impact on the macro-average F1 score. MCC is calculated based on the confusion matrix and considers the overall quantity of samples. Compared to the macro-average F1 score, MCC is less sensitive to different class sample distributions. Therefore, under different train-test splits, MCC exhibits less variation. This further confirms that the insufficient representation of fraudulent samples can lead to poorer performance in identifying smaller class fraudulent instances.

8.6 Conclusion

This chapter designs an online evolving fraud detection framework, named EvoFD, for transaction streams with open category and concept drift. The framework learns in an online fashion from the transaction streams. In the framework, the training and inferring are conducted simultaneously and independently, where the model learns continually from the infinite transaction streams. We devise three effective components for the framework, i.e., replay memory, class anchor, and class-balancing loss, for the framework. The balanced replay memory preserves exemplars for each class to approximate the distributions of observed transactions, which facilitates the consolidation of previous information when learning on new transactions. We progressively update the class anchor for each class to represent the centroid of the class. The representation network is optimized to embed the original feature vectors into the unified normalized space with class anchors. The class-balancing loss is utilized to enhance the model learning in the imbalanced distributions. We conduct extensive experiments to validate the advantages of the proposed framework over the real-world transaction dataset.

In future work, we will explore the extension of our work to more models and applications. Firstly, in recent years, graph-based models have been widely concerned. Utilizing graph data has been proven to be a powerful means of mining deep relationships in fraud detection. We will investigate graph-based models to further improve the representability of the learner. Secondly, with the explosive growth of data scale and the essential concern of privacy security, distributed learning has been a significant technology in the practical application of fraud detection methods. We are interested in the combination of continual learning and distributed learning. Moreover, the labeling of transactions is generally a long and complicated process. The real-time performance of supervised learning-based fraud detection methods might be affected due to the lag of transaction labels. Applying weakly supervised learning to online fraud detection is worth exploring. Finally, we will also focus on developing specific techniques that can effectively handle the challenges associated with class forgetting.

References

1. K. Busse, M. Tahaei, K. Krombholz, E. von Zezschwitz, M. Smith, J. Tian, W. Xu, in *IEEE EuroS&P Workshops 2020, Genoa, Italy, September 7–11, 2020* (2020), pp. 200–209
2. C. Wang, H. Zhu, IEEE Trans. Dependable Secure Comput. **19**(1), 301 (2022)
3. C. Liu, L. Sun, X. Ao, J. Feng, Q. He, H. Yang, in *Proceedings of the ACM SIGKDD, Virtual Event, Singapore, August 14–18, 2021* (2021), pp. 3280–3288
4. X. Zhou, Z. Zhang, L. Wang, P. Wang, in *Proceedings of the IJCNN 2019, Budapest, Hungary, July 14–19, 2019* (2019), pp. 1–7
5. C. Wang, H. Zhu, IEEE Trans. Inf. Forensics Secur. **17**, 2703 (2022)
6. Z. Zhang, Z. Wei, L. Ma, IEEE Trans. Comput. Social Syst. **11**(2), 1589 (2024)

7. A. Agarwal, N.K. Ratha, in *Proceedings of the ACM CIKM, Gold Coast, Queensland, Australia, November 1–5, 2021, CEUR Workshop Proceedings*, ed. by G. Cong, M. Ramanath, vol. 3052
8. J. Jiang, B. Ni, C. Wang, in *Proceedings of the WWW, Virtual Event / Ljubljana, Slovenia, April 19–23, 2021* (2021), pp. 238–246
9. C. Liu, Y. Gao, L. Sun, J. Feng, H. Yang, X. Ao, in *Proceedings of the ACM SIGKDD, Washington, DC, USA, August 14–18, 2022* (2022), pp. 3357–3365
10. Z. Zhang, X. Zhou, X. Zhang, L. Wang, P. Wang, et al., Secur. Commun. Netw. **2018**, 5680264:1–5680264:9 (2018)
11. I. Žliobaitė, A. Bifet, J. Read, B. Pfahringer, G. Holmes, Mach. Learn. **98**, 455 (2015)
12. V. Losing, B. Hammer, H. Wersing, Neurocomputing **275**, 1261 (2018)
13. L. Korycki, B. Krawczyk, in *Proceeding of the IEEE ICDE, Chania, Greece, April 19–22, 2021* (2021), pp. 1068–1079
14. S. Mirzadeh, M. Farajtabar, R. Pascanu, H. Ghasemzadeh, in *Proceeding of the NeurIPS, virtual, December 6–12, 2020* (2020)
15. S. Farquhar, Y. Gal, CoRR **abs/1805.09733** (2018)
16. M. Masana, X. Liu, B. Twardowski, M. Menta, A.D. Bagdanov, J. van de Weijer, CoRR **abs/2010.15277** (2020)
17. Z. Li, D. Hoiem, in *Proceeding of the ECCV, Amsterdam, The Netherlands, October 11–14, 2016*, vol. 9908, pp. 614–629
18. S. Rebuffi, A. Kolesnikov, G. Sperl, C.H. Lampert, in *Proceeding of the IEEE CVPR, Honolulu, HI, USA, July 21–26, 2017*, pp. 5533–5542
19. M.D. Lange, R. Aljundi, M. Masana, S. Parisot, X. Jia, A. Leonardis, G.G. Slabaugh, T. Tuytelaars, IEEE Trans. Pattern Anal. Mach. Intell. **44**(7), 3366 (2022)
20. D.L. Silver, R.E. Mercer, in *Conference of the Canadian Society for Computational Studies of Intelligence, Calgary, Canada, May 27–29, 2002*. Lecture Notes in Computer Science, vol. 2338. *Lecture Notes in Computer Science*, vol. 2338, pp. 90–101
21. S.C.H. Hoi, D. Sahoo, J. Lu, P. Zhao, Neurocomputing **459**, 249 (2021)
22. N. Gunasekara, B. Pfahringer, H.M. Gomes, A. Bifet, in *Proceedings of the IJCAI 2023, 19th–25th August 2023, Macao, SAR, China*, pp. 6628–6637
23. F. Ceschin, M. Botacin, A. Bifet, B. Pfahringer, L.S. Oliveira, H.M. Gomes, A. Grégio, Digital Threats: Res. Pract. **5**(1), 1 (2024)
24. Z. Chen, B. Liu, *Lifelong Machine Learning*, 2nd edn. Synthesis Lectures on Artificial Intelligence and Machine Learning (Morgan & Claypool Publishers, California, 2018)
25. D. Lopez-Paz, M. Ranzato, in *Proceedings of the NIPS 2017, Long Beach, CA, USA, December 4–9, 2017*, pp. 6467–6476
26. D. Rolnick, A. Ahuja, J. Schwarz, T.P. Lillicrap, G. Wayne, in *Proceedings of the NeurIPS, Vancouver, BC, Canada, December 8–14, 2019*, pp. 348–358
27. C. Atkinson, B. McCane, L. Szymanski, A.V. Robins, CoRR **abs/1802.03875** (2018)
28. H. Shin, J.K. Lee, J. Kim, J. Kim, in *Proceeding of the NIPS, Long Beach, CA, USA, December 4–9, 2017*, pp. 2990–2999
29. A.R. Triki, R. Aljundi, M.B. Blaschko, T. Tuytelaars, in *Proceedings of the IEEE ICCV, Venice, Italy, October 22–29, 2017*, pp. 1329–1337
30. J. Kirkpatrick, R. Pascanu, N. Rabinowitz, J. Veness, G. Desjardins, A.A. Rusu, K. Milan, J. Quan, T. Ramalho, A. Grabska-Barwinska, et al., Proc. Natl. Acad. Sci. **114**(13), 3521 (2017)
31. C.V. Nguyen, Y. Li, T.D. Bui, R.E. Turner, in *Proceeding of the ICLR, Vancouver, BC, Canada, April 30–May 3, 2018*
32. R. Aljundi, F. Babiloni, M. Elhoseiny, M. Rohrbach, T. Tuytelaars, in *Proceeding of the ECCV, Munich, Germany, September 8–14, 2018*, vol. 11207, pp. 144–161
33. A.A. Rusu, N.C. Rabinowitz, G. Desjardins, H. Soyer, J. Kirkpatrick, K. Kavukcuoglu, R. Pascanu, R. Hadsell, CoRR **abs/1606.04671** (2016)
34. A. Lee, Y. Zhang, H.M. Gomes, A. Bifet, B. Pfahringer, in *Proceedings of the ACM CIKM 2023, Birmingham, United Kingdom, October 21–25, 2023* (2023), pp. 4038–4042

35. J. Xu, Z. Zhu, in *Proceeding of the NeurIPS, Montréal, Canada, December 3–8, 2018* (2018), pp. 907–916
36. R. Aljundi, P. Chakravarty, T. Tuytelaars, in *Proceeding of the IEEE CVPR, Honolulu, HI, USA, July 21–26, 2017* (2017), pp. 7120–7129
37. A. Mallya, S. Lazebnik, in *Proceeding of the IEEE CVPR, Salt Lake City, UT, USA, June 18–22, 2018* (2018), pp. 7765–7773
38. J. Serrà, D. Suris, M. Miron, A. Karatzoglou, in *Proceeding of the ICML, Stockholmsmässan, Stockholm, Sweden, July 10–15, 2018*, vol. 80, pp. 4555–4564
39. W.W. Cohen, in *Proceeding of the ICML, Tahoe City, California, USA, July 9–12, 1995* (1995), pp. 115–123
40. S. Rosset, U. Murad, E. Neumann, Y. Idan, G. Pinkas, in *Proceeding of the ACM SIGKDD, San Diego, CA, USA, August 15–18, 1999* (1999), pp. 409–413
41. R.W. Brause, T.S. Langsdorf, H. Hepp, in *Proceeding of the IEEE ICTAI, Chicago, Illinois, USA, November 8–10, 1999* (1999), pp. 103–106
42. C. Wang, in *Proceeding of the IJCAI 2020, Virtual Event, January 7–15, 2021* (2021), pp. 4611–4618
43. H. Zhu, C. Wang, IEEE Trans. Inf. Forensics Secur. **18**, 5706 (2023)
44. C. Liu, Q. Zhong, X. Ao, L. Sun, W. Lin, J. Feng, Q. He, J. Tang, in *Proceeding of the ACM SIGKDD, Virtual Event, CA, USA, August 23–27, 2020* (2020), pp. 3035–3043
45. Y. Zhu, D. Xi, B. Song, F. Zhuang, S. Chen, X. Gu, Q. He, in *Proceeding of the WWW, Taipei, China, April 20–24, 2020* (2020), pp. 928–938
46. D. Xi, F. Zhuang, B. Song, Y. Zhu, S. Chen, D. Hong, T. Chen, X. Gu, Q. He, in *Proceeding of the ACM SIGIR, Virtual Event, China, July 25–30, 2020* (2020), pp. 1893–1896
47. S. Yuan, X. Wu, J. Li, A. Lu, in *Proceedings of the CIKM, Singapore, November 06–10, 2017* (2017), pp. 2419–2422
48. D. Cheng, X. Wang, Y. Zhang, L. Zhang, IEEE Trans. Knowl. Data Eng. **34**(8), 3800 (2022)
49. P. Zheng, S. Yuan, X. Wu, in *Proceedings of the AAAI, Honolulu, Hawaii, USA, January 27–February 1, 2019* (2019), pp. 1278–1285
50. B. Hooi, K. Shin, H.A. Song, A. Beutel, N. Shah, C. Faloutsos, ACM Trans. Knowl. Discov. Data **11**(4), 44:1 (2017)
51. D.J.N.J. Soemers, T. Brys, K. Driessens, M.H.M. Winands, A. Nowé, in *Proceedings of the AAAI, New Orleans, Louisiana, USA, February 2–7, 2018* (2018), pp. 7831–7836
52. F. Carcillo, Y.L. Borgne, O. Caelen, G. Bontempi, Int. J. Data Sci. Anal. **5**(4), 285 (2018)
53. F. Carcillo, Y.L. Borgne, O. Caelen, Y. Kessaci, F. Oblé, G. Bontempi, Inf. Sci. **557**, 317 (2021)
54. J. Kim, H. Kim, H. Kim, Appl. Intell. **49**(8), 2842 (2019)
55. S. Shehnepoor, R. Togneri, W. Liu, M. Bennamoun, IEEE Trans. Inf. Forensics Secur. **17**, 280 (2022)
56. Y. Hsu, Y. Liu, Z. Kira, CoRR **abs/1810.12488** (2018)
57. G.M. van de Ven, A.S. Tolias, CoRR **abs/1904.07734** (2019)
58. R. Aljundi, K. Kelchtermans, T. Tuytelaars, in *Proceedings of the IEEE CVPR, Long Beach, CA, USA, June 16–20, 2019* (2019), pp. 11,254–11,263
59. J.S. Vitter, ACM Trans. Math. Softw. **11**(1), 37 (1985)

MIX
Papier aus verantwortungsvollen Quellen
Paper from responsible sources
FSC® C105338

If you have any concerns about our products,
you can contact us on
ProductSafety@springernature.com

In case Publisher is established outside the EU,
the EU authorized representative is:
Springer Nature Customer Service Center GmbH
Europaplatz 3, 69115 Heidelberg, Germany

Printed by Libri Plureos GmbH
in Hamburg, Germany